"十三五"普通高等教育本科规划教材
全国高等院校工业设计专业系列规划教材

工业设计概论
（双语）

主　编　窦金花
副主编　张　磊　张慧姝

内 容 简 介

本书按工业设计发展进程摘取了一些重要的英文文章，包括工业设计理论、设计思维、实践案例等内容。本书对一些文章进行了详细的分析与翻译，提取了重要的专业词汇进行讲解，理论知识丰富，内容翔实。本书内容包括工业社会与现代设计的萌芽、现代主义与工业设计、第三次浪潮与后现代主义、当代工业设计理论与实践等几个部分，核心课文配有专业词汇翻译和文章翻译，为学生准确地进行阅读理解提供参考，同时，在每个单元又提取了一些对工业设计发展具有重要意义的人物和案例等阅读材料，供学生课堂与课后学习。

全书共 60 余篇文章，适合作为全国高等院校设计类专业本科学生的教材。

图书在版编目(CIP)数据

工业设计概论：双语/窦金花主编. —北京： 北京大学出版社，2017.4
(全国高等院校工业设计专业系列规划教材)
ISBN 978-7-301-27933-5

Ⅰ.①工⋯　Ⅱ.①窦⋯　Ⅲ.①工业设计—双语教学—高等学校—教材　Ⅳ.①TB47

中国版本图书馆 CIP 数据核字(2017)第 007123 号

书　　　名	工业设计概论（双语）
	GONGYE SHEJI GAILUN
著作责任者	窦金花　主编
策划编辑	童君鑫
责任编辑	李娉婷
标准书号	ISBN 978-7-301-27933-5
出版发行	北京大学出版社
地　　　址	北京市海淀区成府路 205 号　100871
网　　　址	http://www.pup.cn　新浪微博：@北京大学出版社
电子信箱	pup_6@163.com
电　　　话	邮购部 62752015　发行部 62750672　编辑部 62750667
印刷者	北京鑫海金澳胶印有限公司
经销者	新华书店
	787 毫米×1092 毫米　16 开本　14.25 印张　324 千字
	2017 年 4 月第 1 版　2017 年 4 月第 1 次印刷
定　　　价	35.00 元

未经许可，不得以任何方式复制或抄袭本书之部分或全部内容。
版权所有，侵权必究
举报电话：010-62752024　电子信箱：fd@pup.pku.edu.cn
图书如有印装质量问题，请与出版部联系，电话：010-62756370

前　言

设计教育在我国起步比较晚，但是发展速度非常快，目前规模也很庞大，特别是近十年来，中国已经成为全球规模最大的设计教育大国。随着社会与经济的发展，全球化视野对设计教育的发展提出新的要求。全球化是一个更广泛、更深刻的概念，它是在经济全球化的基础上，以先进的交通工具和通信工具为载体，尤其是在互联网的帮助下，将全世界人们的生活联系成一个有机的整体。在这个背景下，世界各地联系越来越紧密，生产、销售和消费活动在全世界范围内展开，整个世界成了一个大市场，跨国和跨地区生产和销售无处不在。中国自改革开放以来，快速加入了全球市场，在这种全球化的背景下，工业设计人才应该具有前沿的设计理念，全球化的设计视野，以及较强的与世界各国人民沟通交流的能力，这对我国的设计教育提出了新的挑战。

当前，无论是从我国设计教育本身的发展来看，还是从我国经济发展的进程来说，设计师必须具有较强的与国外同行或客户交流的能力，因此，在设计教育中实施双语教学是非常必要的。目前，许多艺术院校还少有课程实行双语教学，只是专门开设了专业英语课程。传统的专业英语文章晦涩难懂，上课方式不够灵活，难以激发学生学习专业外语的兴趣，加之有些学生英语底子比较薄弱，很难理解授课内容，更难自由地运用专业英语知识进行国际学习交流，因此，毕业生在综合素质上还有一定欠缺。目前，设计类相关双语教材比较少，本书的编写符合工业设计教学未来发展趋势与高等教育需求，希望为学生英语能力的提高提供更多的帮助。

本书与同类教材相比，文章难度适中，条理脉络更为清晰，使学生能够结合设计理论知识进行专业英语学习，巩固所学专业知识。同时，本书选取了许多目前流行的社会理念与优秀设计案例，符合工业设计专业不断发展变化的趋势。

感谢我的同事张磊老师为本书的编写尽心尽力；感谢北京联合大学的张慧姝老师为本书的编写提供了有价值的指导和建议；因为有大家的共同努力，本书才能顺利完成。

窦金花

2017 年 2 月

目　　录

Unit One　Industrial Society and Sprout of Modern Design ······ 1

　　Core Text 1　Industrial Society ······ 2
　　Core Text 2　Arts and Crafts Movement ······ 11
　　Core Text 3　Art Nouveau ······ 23

Unit Two　Modernism and Industrial Design ······ 33

　　Core Text 4　Constructivism ······ 34
　　Core Text 5　Modernism ······ 41
　　Core Text 6　Bauhaus ······ 57
　　Core Text 7　Raymond Loewy ······ 73

Unit Three　The Third Wave and Post-Modernism ······ 91

　　Core Text 8　Development of the Information Society Model ······ 92
　　Core Text 9　Pop Design ······ 104

Unit Four　Contemporary Industrial Design Theory and Practice ······ 119

　　Core Text 10　Industrial Design ······ 120
　　Core Text 11　The Nature of Design ······ 134
　　Core Text 12　Design Management ······ 163
　　Core Text 13　IDEO ······ 185
　　Core Text 14　Responsible Products: Selecting Design and Materials ······ 203

参考文献 ······ 215

Unit One
Industrial Society and Sprout of Modern Design

Core Text 1

Industrial Society

Industrial society: A society which exhibits an extended division of labour and a reliance on large-scale production using power-driven machinery. This characterization does not include any specification about markets, and thus industrial society has been seen as a common designation for recent capitalist and socialist formations. Saint Simon, who used the category of industrial society in historical contrast with military society, envisaged a technocratic future. Other writers who were conscious of the emergence of a new form of market society emphasized a further characteristic: widespread participation in the labour market, coupled with very limited participation of the direct producers in the product market. Marx, for example, saw this as one characteristic of the capitalist form of industrial society. It has been suggested that post-industrial society has now emerged. In post-industrial society, division of labour may be looser than in industrial society because people have transferable skills; accordingly, the industrial discipline of fordism is looser as well. Hence some Marxist scholars call modern post-industrial societies "post-fordist".

In sociology, industrial society refers to a society with a modern societal structure. Such a structure developed in the west in the period of time following the industrial revolution. Pre-modern, or Pre-industrial society are also called agrarian societies. Industrial societies are generally mass societies.

Industrial society is characterized by the use of external energy sources, such as fossil fuels, to increase the rate and scale of production. The production of food is shifted to large commercial farms where the products of industry, such as combine harvesters and petroleum based fertilizers, are used to decrease required human labor while increasing production. No longer needed for the production of food, excess labor is moved into these factories where mechanization is utilized to further increase efficiency. As populations grow, and mechanization is further refined, often to the level of automation, many workers shift to expanding service industries.

Industrial society makes urbanization desirable, in part so that workers can be closer to centers of production, and the service industry can provide labor to workers and those that benefit financially from them, in exchange for a piece of production profits with which they can buy goods. This leads to the rise of very large cities and surrounding suburban areas with a high rate of economic activity.

These urban centers require the input of external energy sources in order to overcome the diminishing returns of agricultural consolidation, due partially to the lack of nearby arable land, associated transportation and storage costs, and are otherwise unsustainable. This makes the reliable availability of the needed energy resources high priority in industrial government policies.

Some theoreticians—namely Ulrich Beck, Anthony Giddens and Manuel Castells—argue that we are located in the middle of a transformation or transition from industrial societies to post-modern societies. The triggering technology for the change from an agricultural to an industrial organisation was steam power, allowing mass production and reducing the agricultural work necessary. Thus many industrial cities are built around rivers. Identified as catalyst or trigger for the transition to post-modern or informational society is global information technology.

Key Words

[1] division [dɪˈvɪʒən] n. 1. 分开；分配 2. 部门 3. 界限
[2] labour [ˈleɪbə] n. 1. 劳动；工作 2. 劳工，工人
[3] industrial society 工业社会
[4] sociology [ˌsəʊsɪˈɒlədʒɪ] n. 社会学
[5] automation [ˌɔːtəˈmeɪʃən] n. 自动化（技术），自动操作
[6] consolidation [kənˌsɒlɪˈdeɪʃən] n. 1. 巩固，加强，强化 2. 联合，统一；合并 3. 变坚固
[7] availability [əˌveɪləˈbɪlɪtɪ] n. 1. 有效；有益；可利用性 2. 可得到的东西（或人）；可得性

Key Sentences

In sociology, industrial society refers to a society with a modern societal structure. Such a structure developed in the west in the period of time following the industrial revolution. Pre-modern, or Pre-industrial society are also called agrarian societies. Industrial societies are generally mass societies.

在社会学中，工业社会是指一种具有现代社会结构的社会。这种结构发展于西方工业革命之后。前现代，前工业社会也同样被称作农业社会。工业社会通常是指大众社会。

课文翻译

工 业 社 会

工业社会是展示劳动的进一步分工和依赖于使用动力驱动机器进行大规模生产的社会。其特征不包括任何市场规范，因此，工业社会已经被认为是近代资本主义和社会主义形式的一个公用符号。圣·西蒙用历史上的工业社会类别与军事社会做对比，预想了一个科技性的未来。其他一些意识到市场社会新形式兴起的作家，也强调了其近一步的特征：劳动市场的广泛分布，以及商品市场中直接生产商的有限加入。例如，马克思认识到这是工业社会资本主义形式的一种特征，这些暗示着后工业社会已经崭露头角。在后工业社会中，劳动分配比工业社会更宽松，因为人们具有可转换的技能；因此，福特制的工业规范也更宽松，因此，一些马克思主义学者称现代后工业社会为"后福特主义时期"。

在社会学中，工业社会是指一个具有现代社会结构的社会，这种结构发展于西方工业革命之后。前现代，前工业社会也同样被称作农业社会。工业社会通常是指大众社会。

工业社会的特点是使用外部能源，如矿物燃料，来提高生产速度和生产规模。食品生产转移到大型商业农场，那里的工业产品，如联合收割机和石油基肥料，可用于减少所需的人力劳动，同时提高产量。剩余劳动力不再为食品生产所需，而被转移到那些利用机械化以进一步提高效率的工厂。人口的增加以及机械化的进一步完善，提高了自动化水平，许多工人转向扩大的服务行业。

工业社会使城市理想化，在某种程度上，工人们可以更接近生产中心，并且服务性行业能够为工人们提供工作，使其获得经济利益，工人们可用换取的单件生产效益来购买商品。这导致了大型城市的增长和周边郊区的经济活动大幅上升。

这些城市中心需要外部资源的投入，以克服农业合并的收益递减，部分是由于附近耕地缺乏，相关的运输、储存等费用，以及其他一些不可持续的因素。这使得可靠的获取所需的资源在政府工业政策中具有优先地位。

一些理论家——乌尔里希·贝克，安东尼·吉登斯与曼纽尔卡斯特—主张我们正处于一个变革的中心，是从工业社会到后现代社会的过渡时期。从农业向工业组织转变的启动技术是蒸汽动力，允许进行大规模生产，降低必需的农业工作。因此，许多工业城市均绕河建立。全球信息技术被认为是过渡到后现代社会或信息化社会的催化剂或触发器。

Free Reading 1

The Architecture of Industrial Civilization—the Second Wave

Three hundred years ago, give or take a half-century, an explosion was heard that sent concussive shock waves racing across the earth, demolishing ancient societies and creating a wholly new civilization. This explosion was, of course, the industrial revolution. And the giant tidal force is set loose on the world—the Second Wave—collided with all the institutions of the past and changed the way of life of millions.

Industrial revolution was launched the Second Wave and creating a strange, powerful, feverishly energetic counter civilization. Industrialism was more than smokestacks and assembly lines. It was a rich, many-sided social system that touched every aspect of human life and attacked every feature of the First Wave past. It produced the great Willow Run factory outside Detroit, but it also put the tractor on the farm, the typewriter in the office, the refrigerator in the kitchen. It produced the daily newspaper and the cinema, the subway and the DC-3. It gave us cubism and twelve-tone music. It gave us Bauhaus buildings and Barcelona chairs, sit-down strikes, vitamin pills, and lengthened life spans. It universalized the wristwatch and the ballot box. More important, it linked all these things together—assembled them, like a machine—to form the most powerful, cohesive and expansive social system the world had ever know: Second Wave civilization.

The precondition of any civilization, old or new, is energy. First Wave societies drew their energy from "living batteries" —human and animal muscle-power—or from sun, wind, and water. Forests were cut for cooking and heating. Waterwheels, some of them using tidal power, turned millstones. Windmills creaked in the fields. Animals pulled the plow. As late as the French Revolution, it has been estimated, Europe drew energy from an estimated 14 million horses and 24 million oxen. All First Wave societies thus exploited energy sources that were renewable. Nature could eventually replenish the forests they cut, the wind that filled their sails, the rivers that turned then paddle wheels. Even animals and people were replaceable "energy slaves".

All Second Wave societies, by contrast, began to draw their energy from coal, gas, and oil—from irreplaceable fossil fuels. This revolutionary shift, coming after Newcomen invented a workable steam engine in 1712, meant that for the first time a civilization was eating into nature's capital rather than merely living off the interest it provided.

This dipping into the earth's energy reserves provided a hidden subsidy for industrial civilization, vastly accelerating its economic growth. And from that day to this, wherever the Second Wave passed, nations built towering technological and economic structures on the assumption that cheap fossil fuels would be endlessly available. In capitalist and communist industrial societies alike, in East and West, this same shift has been apparent—from dispersed to concentrated energy, from renewable to non-renewable, from many different sources and fuels to a few. Fossil fuels formed the energy base of all Second Wave societies.

The leap to a new energy system was paralleled by a gigantic advance in technology. First Wave societies had relied on what Vitruvius, two thousand years ago, called "necessary inventions". But these early winches and wedges, catapults, winepresses, levers, and hoists were chiefly used to amplify human or animal muscles.

The Second Wave pushed technology to a totally new level. It spawned gigantic electromechanical machines, moving parts, belts, hoses, bearings, and bolts—all clattering and ratcheting along. And these new machines did more than augment raw muscle. Industrial civilization gave technology sensory organs, creating machines that could hear, see, and touch with greater accuracy and precision than human beings. It gave technology a womb, by inventing machines designed to give birth to new machines in infinite progression—i.e., machine tools. More important, it brought machines together in interconnected systems under a single roof, to create the factory and ultimately the assembly line within the factory.

On this technological base a host of industries sprang up to give Second Wave civilization its defining stamp. At first there were coal, textiles, and railroads, then steel, auto manufacture, aluminum, chemicals, and appliances. Huge factory cities leaped into existence: Lille and Manchester for textiles, Detroit for automobiles, Essen and—later—Magnitogorsk for steel and a hundred others as well.

From these industrial centers poured millions upon endless millions of identical products—shirts, shoes, automobiles, watches, toys, soap, shampoo, cameras, machine guns, and electric motors. The new technology powered by the new energy system opened the door to mass production.

The Second Wave wrought changes in this creaking, overburdened distribution system that were as radical, in their ways, as the more publicized advances made in production.

Railroads, highways, and canals opened up the hinterlands, and with industrialism came "palaces of trade" —the first department stores. Complex networks of jobbers, wholesalers, commission agents, and manufacturers' representatives sprang up, and in 1871 George Huntington Hartford, whose first store in New York was painted vermilion and had a cashier's cage shaped like a Chinese pagoda, did for distribution what Henry Ford later did for the factory. He advanced it to an entirely new stage by creating the world's first mammoth chain-store system—The Great Atlantic and Pacific Tea Company.

Custom distribution gave way to the mass distribution and mass merchandising that became as familiar and central a component of all industrial societies as the machine itself.

What we see, therefore, if we take these changes together, is a transformation of what might be called the "techno-sphere". All societies—primitive, agricultural, or industrial—use energy; they make things; they distribute things. In all societies the energy system, the production system, and the distribution system are interrelated parts of something larger. This larger system is the techno-sphere, and it has a characteristic form at each stage of social development.

As the Second Wave swept across the planet, the agricultural techno-sphere was replaced by an industrial techno-sphere: non-renewable energies were directly plugged into a mass production system which, in turn, spewed goods into a highly developed mass distribution system.

This Second Wave techno-sphere, however, needed an equally revolutionary "sociosphere" to accommodate it. It needed radically new forms of social organization.

Before the industrial revolution, for example, family forms varied from place to place. But wherever agriculture held sway, people tended to live in large, multigenerational households, with uncles, aunts, in-laws, grandparents, or cousins all living under the same roof, all working together as an economic production unit—from the "joint family" in India to the "zadruga" in the Balkans and the "extended family" in Western Europe. And the family was immobile—rooted to the soil.

As the Second Wave began to move across First Wave societies, families felt the stress of change. Within each household the collision of wave fronts took the form of conflict, attacks on patriarchal authority, altered relationships between children and parents, new notions of propriety. As economic production shifted from the field to the factory, the family no longer worked together as a unit. To free workers for factory labor, key functions of the family were parceled out to new, specialized institutions. Education of the child was turned over to schools. Care of the aged was turned over to poor-houses or old-age homes or nursing homes. Above all, the new society required mobility. It needed workers who would follow jobs from place to place.

Burdened with elderly relatives, the sick, the handicapped, and a large brood of children, the extended family was anything but mobile. Gradually and painfully, therefore, family structure began to change. Tom apart by the migration to the cities, battered by economic storms, families stripped themselves of unwanted relatives, grew smaller, more mobile, and more suited to the needs of the new techno-sphere.

The so-called nuclear family—father, mother, and a few children, with no encumbering relatives—became the standard, socially approved, "modern" model in all industrial societies, whether capitalist or socialist. Even in Japan, where ancestor worship gave the elderly an exceptionally important role, the large, close-knit, multigenerational household began to break down as the Second Wave advanced. More and more nuclear units appeared. In short, the nuclear family became an identifiable feature of all Second Wave societies, marking them off from First Wave societies just as surely as fossil fuels, steel mills, or chain stores.

As work shifted out of the fields and the home, moreover, children had to be prepared for factory life. The early mine, mill, and factory owners of industrializing England discovered, as Andrew Ure wrote in 1835, that it was "nearly impossible to convert persons past the age of puberty, whether drawn from rural or from handicraft occupations, into useful factory hands." If young people could be prefitted to the industrial system, it would vastly ease the problems of industrial discipline later on. The result was another central structure of all Second Wave societies: mass education.

Built on the factory model, mass education taught basic reading, writing, and arithmetic, a bit of history and other subjects. This was the "overt curriculum". But beneath it lay an invisible or "covert curriculum" that was far more basic. It consisted—and still does in most industrial nations—of three courses: one in punctuality, one in obedience, and one in rote, repetitive work. Factory labor demanded workers who showed up on time, especially assembly-line hands. It demanded workers who would take orders from a management hierarchy without questioning. And it demanded men and women prepared to slave away at machines or in offices, performing brutally repetitious operations.

Thus from the mid-nineteenth century on, as the Second Wave cut across country after country, one found a relentless educational progression: children started school at a younger and younger age, the school year became longer and longer (in the United States it climbed 35 percent between 1878 and 1956), and the number of years of compulsory schooling irresistibly increased.

Mass public education was clearly a humanizing step forward. As a group of mechanics and workingmen in New York City declared in 1829, "Next to life and liberty, we consider education the greatest blessing bestowed upon mankind." Nevertheless, Second Wave schools machined generation after generation of young people into a pliable, regimented work force of the type required by electromechanical technology and the assembly line.

All human groups, from primitive times to today, depend on face-to-face, person-to-person communication. But systems were needed for sending messages across time and space as well. The ancient Persians are said to have set up towers or "call-posts", placing men with shrill, loud voices atop them to relay messages by shouting from one tower to the next. The Romans operated an extensive messenger service called the cursus publicus.

Between 1305 and the early 1800's, the House of Taxis ran a form of pony express service all over Europe. By 1628 it employed twenty thousand men. Its couriers, clad in blue and silver uniforms, crisscrossed the continent carrying messages between princes and generals, merchants and money lenders.

The information needed for economic production in primitive and First Wave societies is comparatively simple and usually available from someone near at hand. It is mostly oral or gestural hi form. Second Wave economies, by contrast, required the tight coordination of work done at many locations. Not only raw materials but great amounts of information had to be produced and carefully distributed.

For this reason, as the Second Wave gained momentum every country raced to build a postal service. The post office was an invention quite as imaginative and socially useful as the cotton gin or the spinning jenny and, to an extent forgotten today, it elicited rhapsodic enthusiasm. The American orator Edward Everett declared, "I am compelled to regard the Post-office, next to Christianity, as the right arm of our modern civilization."

Nor could the mushrooming informational needs of industrial societies be met in writing alone. Thus the telephone and telegraph were invented in the nineteenth century to carry then: share of the ever swelling communications load. By 1960 Americans were placing some 256 million phone calls per day—over 93 billion a year—and even the most advanced telephone systems and networks in the world were often overloaded.

Today, of course, the mass circulation newspaper and magazine are so standard a part of daily life in every one of the industrial nations that they are taken for granted. Yet the rise of these publications on a national level reflected the convergent development of many new industrial technologies and social forms. Thus, writes Jean-Louis Servan-Schreiber, they were made possible by the coming together of "trains to transport the publications throughout a European-size country in a single day; rotary presses capable of turning out dozens of millions of copies in several hours; a network of telegraph and telephones ... above all a public taught to read by compulsory education, and industries needing to mass distribute their products."

In the mass media, from newspapers and radio to movies and television, we find once again an embodiment of the basic principle of the factory. All of them stamp identical messages into millions of brains, just as the factory stamps out identical products for use in millions of homes. Standardized, mass-manufactured "facts", counterparts of standardized, mass-manufactured products, flow from a few concentrated image-factories out to millions of consumers. Without this vast, powerful system for channeling information, industrial civilization could not have taken form or functioned reliably.

The Second Wave brought with it a fantastic extension of human hope. For the first time men and women dared to believe that poverty, hunger, disease, and tyranny might be overthrown. Utopian writers and philosophers, from Abbe Morelly and Robert Owen to Saint-Simon, Fourier, Proudhon, Louis Blanc, Edward Bellamy, and scores of others, saw in the emerging industrial civilization the potential for introducing peace, harmony, employment for all, equality of wealth or of opportunity, the end of privilege based on birth, the end of all those conditions that seemed immutable or eternal during the hundreds of thousands of years of primitive existence and the thousands of years of agricultural civilization.

Free Reading 2

Standardization

The most familiar of these Second Wave principles is standardization. Everyone knows that industrial societies turn out millions of identical products. Fewer people have stopped to notice, however, that once the market became important, we did more than simply standardize Coca-Cola bottles, light bulbs, and auto transmissions. We applied the same principle to many other things. Among the first to grasp the importance of this idea was Theodore Vail who, at the turn of the century, built the American Telephone & Telegram Company (AT&T) into a giant.

Working as a railway postal clerk in the late 1860's, Vail had noticed that no two letters necessarily went to their destinations via the same route. Sacks of mail traveled back and forth, often taking weeks or months to reach their destinations. Vail introduced the idea of standardized routing—all letters going to the same place would go the same way—and helped revolutionize the post office. When he later formed AT&T, he set out to place an identical telephone in every American home.

Vail standardized not only the telephone handset and all its components but AT&T's business procedures and administration as well. In a 1908 advertisement he justified his swallowing up small telephone companies by arguing for "a clearing-house of standardization" that would ensure economy in "construction of equipment, lines and conduits, as well as the operating methods and legal work," not to mention "a uniform system of operating and accounting." What Vail recognized is that to succeed in the Second Wave environment, "software" —i. e., procedures and administrative routines—had to be standardized along with hardware.

Vail was only one of the great standardizers who shaped industrial society. Another was Frederick Winslow Taylor, a machinist turned crusader, who believed that work could be made scientific by standardizing the steps each worker performed. In the early decades of this century Taylor decided that there was one best (standard) way to perform each job, one best (standard) tool to perform it with, and a stipulated (standard) tune which to complete it. Armed with this philosophy, he became the world's leading management guru. In his time, and later, he was compared with Freud, Marx, and Franklin.

In Second Wave societies, hiring procedures as well as work were increasingly standardized. Standardized tests were used to identify and weed out the supposedly unfit, especially in the civil service. Pay scales were standardized throughout whole industries, along with fringe benefits, lunch hours, holidays, and grievance procedures. To prepare youth for the job market, educators designed standardized curricula. Men like Binet and Terman devised standardized intelligence tests. School grading policies, admission procedures, and accreditation rules were similarly standardized.

The mass media, meanwhile, disseminated standardizing imagery, so that millions read the same advertisements, the same news, and the same short stories. The repression of minority languages by central governments, combined with the influence of mass communications, led to the near disappearance of local and regional dialects or even whole languages, such as Welsh and Alsatian. "Standard" American, English, French, or, for that matter, Russian, supplanted "nonstandard" languages. Different parts of the country began to look alike, as identical gas stations, billboards, and houses cropped up everywhere. The principle of standardization ran through every aspect of daily life. At an even deeper level, industrial civilization needed standardized weights and measures. It is no accident that one of the first acts of the French Revolution, which ushered the age of industrialism into France.

Moreover, if mass production required the standardization of machines, products, and processes, the ever-expanding market demanded a corresponding standardization of money, and even prices. Historically, money had been issued by banks and private individuals as well as by kings. Even as late as the 19 century privately minted money was still in use in parts of the United States, and the practice lasted until 1935 in Canada. Gradually, however, industrializing nations suppressed all nongovernmental currencies and managed to impose a single standard currency in their place.

In 1825 a young Northern Irish immigrant named A. T. Stewart arrived in New York, opened a dry goods store, and shocked customers and competitors alike by introducing a fixed price for every item. This one-price policy—price standardization—made Stewart one of the merchant princes of his era and cleared away one of the key obstacles to the development of mass distribution.

Whatever their other disagreements, advanced Second Wave thinkers shared the conviction that standardization was efficient.

Core Text 2

Arts and Crafts Movement

The Arts and Crafts Movement was a British, Canadian, Australian, and American aesthetic movement occurring in the last years of the 19th century and the early years of the 20th century. Inspired by the writings of John Ruskin and a romantic idealization of a craftsperson taking pride in their personal handiwork, it was at its height between approximately 1880 and 1910. drawing its support from progressive artists, architects and designers, philanthropists, amateurs and middle-class women seeking work in the home. They set up small workshops apart from the world of industry, revived old techniques and revered the humble household objects of pre-industrial times.

The ideology of the Arts and Crafts Movement represented a reaction against the moral and material consequences of the industrial revolution. Its followers were concerned with the negative social and aesthetic impact of Victorian urbanization and what was believed to be an assault on the creative integrity of the design process through the division of labour and other industrial methods of production. Among the major designers associated with the movement were William Morris, A. H. Mackmurdo, Lewis F. Day, C. R. Ashbee, and C. F. Voysey. The roots of the movement lay in the writings and work of the architect and designer Augustus Welby Pugin (1812-1852) and the eloquent Victorian artist and critic John Ruskin. The latter's major books *The Seven Lamps of Architecture* (1849) and *The Stones of Venice* (1851) equate the quality of design with the quality of the society that produced it, drawing analogies between the decline and fall of the Venetian Empire and social and aesthetic change in Victorian Britain. He called for a rejection of the increasing material preoccupations of contemporary society and a return to the dignity of labour enjoyed in pre-industrial times. William Morris, a dominant figure in the Arts and Crafts Movement, explored similar ideas in his writings and design work.

Arts and Crafts designers embraced a number of common principles. These included an honest use of materials and methods of construction as opposed to the more widespread celebration of ingenious applications of new materials and processes to imitate other production processes and finishes. There was also a widespread use of nature-based decorative motifs and a general commitment to the principles of craft, rather than industrialized production. There was also a move by some groups of designers away from towns and cities to rural locations as a means of creating communities of craft workers, often building on the vernacular skills still being practised in the country.

Typifying such an outlook was Charles Robert Ashbee's Guild of Handicraft, which moved to Chipping Camden in Gloucestershire in 1902. From the 1880s onwards a number of Guilds were established including the Art Workers' Guild (established 1884), the Century Guild (1882), and the Guild of Handicraft (established 1888). The idea of Guilds looked back to medieval times when groups of craftsmen worked together collaboratively. An early Arts and craft manifestation of this idea had been William Morris's firm, Morris, Marshall, Faulkner, and Co. (See Morris & Co.) founded in London in 1861. Many of the ideas of the Arts and Crafts Movement were disseminated by the writings of a number of its participants, magazines such as the Century Guild's *Hobby Horse* (established 1884) or the *Studio* (established 1893), and participation in exhibitions at home and abroad.

1. Key Principles

Yet, while the Arts and Crafts movement was in large part a reaction to industrialization, if looked at on the whole, it was neither anti-industrial nor anti-modern. Some of the European factions believed that machines were in fact necessary, but they should only be used to relieve the tedium of mundane, repetitive tasks. At the same time, some Arts and Crafts leaders felt that objects should also be affordable. The conflict between quality production and "demo" design, and the attempt to reconcile the two, dominated design debate at the turn of the twentieth century.

Those who sought compromise between the efficiency of the machine and the skill of the craftsman thought it a useful endeavour to seek the means through which a true craftsman could master a machine to do his bidding, in opposition to what many believed to be the reality during the Industrial Age, i.e., that humans had become slaves to the industrial machine.

The need to reverse the human subservience to the unquenchable machine was a point that everyone agreed on. Yet the extent to which the machine was ostracised from the process was a point of contention debated by many different factions within the Arts and Crafts movement throughout Europe.

This conflict was exemplified in the German Arts and Crafts movement, by the clash between two leading figures of the Deutscher Werkbund (DWB), Hermann Muthesius and Henry Van de Velde. Muthesius, also head of design education for German Government, was a champion of standardization. He believed in mass production, in affordable democratic art. Van de Velde, on the other hand, saw mass production as a threat to creativity and individuality.

Though the spontaneous personality of the designer became more central than the historical "style" of a design, certain tendencies stood out: reformist neo-gothic influences, rustic and "cottagey" surfaces, repeating designs, vertical and elongated forms. In order to express the beauty inherent in craft, some products were deliberately left slightly unfinished, resulting in a certain rustic and robust effect. There were also socialist undertones to this movement—most explicitly, and primarily, in Great Britain—in that another primary aim was for craftspeople to derive satisfaction from what they did. This satisfaction, the proponents of this movement felt, was totally denied in the industrialised processes inherent in compartmentalised machine production.

In fact, the proponents of the Arts and Crafts movement were against the principle of a division of labour, which in some cases could be independent of the presence or absence of machines. They were in favour of the idea of the master craftsman, creating all the parts of an item of furniture, for instance, and also taking a part in its assembly and finishing, with some possible help by apprentices. This was in contrast to work environments such as the French Manufactories, where everything was oriented towards the fastest production possible. (For example, one person or team would handle all the legs of a piece of furniture, another all the panels, another assembled the parts and yet another painted and varnished or handled other finishing work, all according to a plan laid out by a furniture designer who would never actually work on the item during its creation.) The Arts and Crafts movement sought to reunite what had been ripped asunder in the nature of human work, having the designer work with his hands at every step of creation. Some of the most famous apostles of the movement, such as Morris, were more than willing to design products for machine production, when this did not involve the wretched division of labour and loss of craft talent, which they denounced. Morris designed numerous carpets for machine production in series.

2. Influences on Later Art

1) Europe

Widely exhibited in Europe, the Arts and Crafts movement's qualities of simplicity and honest use of materials negating historicism inspired designers like Henry van de Velde and movements such as Art Nouveau, the Dutch De Stijl group, Vienna Secession, and eventually the Bauhaus. The movement can be assessed as a prelude to Modernism, where pure forms, stripped of historical associations, would be once again applied to industrial production.

In Russia, Viktor Hartmann, Viktor Vasnetsov and other artists associated with Abramtsevo Colony sought to revive the spirit and quality of medieval Russian decorative arts in the movement quite independent from that flourishing in Great Britain.

The Wiener Werkstätte, founded in 1903 by Josef Hoffmann and Koloman Moser, played an independent role in the development of Modernism, with its Wiener Werkstätte Style.

The British Utility furniture of World War II was simple in design and based on Arts and Crafts ideas.

In Ireland, the Honan Chapel, located in Cork, Ireland, on the grounds of University College Cork, built in 1916 is internationally recognised as representative of the Irish Arts and Crafts movement, as shown in Figure 1.1.

2) United States

In the United States, the Arts and Crafts Movement took on a distinctively more bourgeois flavor. While the European movement tried to recreate the virtuous world of craft labor that was being destroyed by industrialization, Americans tried to establish a new source of virtue to replace heroic craft production: the tasteful middle-class home. They thought that the simple but refined aesthetics of Arts and Crafts decorative arts would ennoble the new experience of industrial consumerism, making individuals more

Figure 1.1　Honan Chapel

rational and society more harmonious. In short, the American Arts and Crafts Movement was the aesthetic counterpart of its contemporary political movement: Progressivism.

In the United States, the Arts and Crafts Movement spawned a wide variety of attempts to reinterpret European Arts and Crafts ideals for Americans. These included the "Craftsman"-style architecture, furniture, and other decorative arts such as the designs promoted by Gustav Stickley in his magazine, *The Craftsman*. A host of imitators of Stickley's furniture (the designs of which are often mislabeled the "Mission Style") included three companies formed by his brothers, the Roycroft community founded by Elbert Hubbard, the "Prairie School" of Frank Lloyd Wright, the Country Day School movement, the bungalow style of houses popularized by Greene and Greene, utopian communities like Byrdcliffe and Rose Valley, developments such as Mountain Lakes, New Jersey, featuring clusters of bungalow and chateau homes built by Herbert J. Hapgood, and the contemporary studio craft movement. Studio pottery—exemplified by Grueby, Newcomb, Teco, Overbeck and Rookwood pottery, Bernard Leach in Britain, and Mary Chase Perry Stratton's Pewabic Pottery in Detroit—as well as the art tiles by Ernest A. Batchelder in Pasadena, California, and idiosyncratic furniture of Charles Rohlfs also demonstrate the clear influence of Arts and Crafts Movement. Mission, Prairie, and the "California bungalow" styles of homebuilding remain tremendously popular in the United States today.

Key Words

[1] Arts and Crafts Movement 工艺美术运动

[2] ideology [ˌaɪdɪˈɒlədʒɪ] *n.* 1. 思想(体系),思想意识 2. 意识形态;观念形态

[3] integrity [ɪnˈtegrɪtɪ] *n.* 1. 正直;诚实,诚恳 2. 完整,完全,完善 3. 廉正 4. 健全

[4] equate [ɪˈkweɪt] *vt.* 1. 认为某事物(与另一事物)相等或相仿 2. 相当于;等于 3. 把(一事物)和(另一事物)等同看待

[5] dignity [ˈdɪgnɪtɪ] n. 1. 庄严，端庄，尊严 2. 高尚，尊贵，高贵 3. 自豪；自尊；自重

[6] ingenious [ɪnˈdʒiːnɪəs] adj. 1. 灵巧的，善于创造发明的 2. 设计独特的，制作精巧的，新颖独特的，巧妙的

[7] disseminate [dɪˈsemɪneɪt] vt. 散布，传播（信息、知识等）

[8] artisan [ˌɑːtɪˈzæn] n. 技工；工匠

[9] tedium [ˈtiːdɪəm] n. 单调乏味，令人生厌，冗长

[10] mundane [mʌnˈdeɪn, ˈmʌndeɪn] adj. 1. 平凡的，平淡的 2. 单调的 3. 世俗的

[11] endeavour [ɪnˈdevə] n. 尽力，竭力 vi. 努力；尽力；竭力

[12] unquenchable [ʌnˈkwentʃəbəl] adj. 难抑制的，不能消灭的

[13] spontaneous [spɒnˈteɪnɪəs] adj. 自发的；自然的；天然产生的；无意识的

[14] bourgeois [ˈbʊəʒwɑː] adj. 资产阶级的 n. 资产阶级分子

[15] progressivism [prəʊˈgresɪvɪəm] n. 进步主义，革新论

Key Sentences

1. Its followers were concerned with the negative social and aesthetic impact of Victorian urbanization and what was believed to be an assault on the creative integrity of the design process through the division of labour and other industrial methods of production.

它的追随者们担心维多利亚城市化带来消极的社会和审美影响，以及被认为是通过分工和其他工业生产方法来攻击设计过程的创造性和完整性。

2. The need to reverse the human subservience to the unquenchable machine was a point that everyone agreed on. Yet the extent to which the machine was ostracised from the process was a point of contention debated by many different factions within the Arts and Crafts movement throughout Europe.

扭转人类对难以遏制的机器信奉的需要是人人都同意的重要问题。然而，机器应该被排斥到何种程度是欧洲各个工艺美术运动派系争论的问题。

3. Though the spontaneous personality of the designer became more central than the historical "style" of a design, certain tendencies stood out：reformist neo-gothic influences，rustic and "cottagey" surfaces, repeating designs, vertical and elongated forms.

虽然设计师自发的个性比起设计的历史风格变得更加重要，某些倾向脱颖而出：改革新哥特式影响、乡村和村舍式的外观、重复设计、纵向和拉长形式。

课文翻译

工艺美术运动

工艺美术运动是19世纪末期到20世纪初期发生在英国、加拿大、澳大利亚和美国的一场艺术运动。它受约翰·拉斯金(John Ruskin)的著作和工匠对其个人手艺高度自豪的浪漫理想化思想的激励，在大约1880年到1910年间处于鼎盛时期。它的支持者包括一些进步的艺术家、建筑师和设计师、慈善家、业余爱好者和希望在家中工作的中产阶级妇

女。他们建立了工业世界之外的小作坊，复兴传统技术，崇尚工业社会之前朴实的家庭用品。

工艺美术运动的思想体系表现了对工业革命道德和物质成果的反对。它的追随者们担心维多利亚城市化带来的消极的社会和审美影响，以及被认为是通过分工和其他工业生产方法来攻击设计过程的创造性和完整性。与这场运动有关的设计师主要是威廉·莫里斯，A. H. 玛克穆多，刘易斯·F. 台，查尔斯·罗伯特·阿什比和 C.F. 沃伊齐。运动起源于建筑师和设计师奥古斯都·韦尔比·普金以及雄辩的维多利亚时代艺术家和评论家约翰·拉斯金的文章与著作。后者的主要著作《建筑的七盏灯》（1849）和《威尼斯的石头》把设计质量和社会质量视作等同，类比了威尼斯帝国的衰落和维多利亚时代英国社会的审美变化。他呼吁抵制当代社会对物质增长的专注，回归到前工业化时代劳动受尊敬的状态。工艺美术运动的主导人物威廉·莫里斯在他的文章和著作中表达了类似的思想。

工艺美术家接受了一些共同的原则，包括诚实地使用材料和施工方法，而不是更广泛庆祝新材料的巧妙应用和模仿其他生产工艺和成品的过程；广泛使用基于自然的装饰主题和工艺原则，而不是工业化生产。另外，还有些由设计师组成的团体将远离城镇去农村地区作为一种创建手工艺人团体的手段，这些团体往往建立在仍在使用本地技能的村庄。

这种观点的典型是查尔斯·罗伯特·阿什比的手工业协会，该协会于 1902 年搬到格洛斯特郡的奇平卡姆登。从 19 世纪 80 年代起，一些行业协会建立起来，包括艺术工作者协会（1884 年成立）、世纪协会（1882）和手工业协会（1888 年成立）。当群体工匠一起协作时，行业协会的思想回到了中世纪时期。早期体现这一思想的艺术和工艺表现是威廉·莫里斯的公司——于 1861 年在伦敦成立的"莫里斯，马歇尔，福克纳商行"。工艺美术运动的许多想法被众多参与者的著作传播开来，例如，世纪协会的季刊《玩具马》（*Hobby House*）（1893 年创立）或《工作室》（*Studio*）（1893 年创立），还有国内外展览的参与者。

1. 关键原则

然而，尽管工艺美术运动在很大程度上反工业化，如果从整体上看，它既不反工业化，也不反现代化。一些欧洲派系认为，实际上，机器是必要的，但它们应该只被用来减轻厌烦的单调乏味的重复任务。与此同时，一些艺术手工艺领袖认为，物品应该是负担得起的。高质量生产和"示范"设计之间的冲突和对两者调和的尝试导致了 20 世纪之交的辩论。

那些寻求机器效率和工匠技能互相折中的人认为，寻找让真正的工匠可以掌握机器执行其命令的方法可有效地反对许多人认定已成为现实的工业时代，也就是人类已成为工业机器奴隶的时代。

扭转人类难以遏制的对机器信奉是人人都同意的重要问题。然而，机器应该被排斥到何种程度是欧洲各个工艺美术运动派系争论的问题。

这场冲突的一个例子是在德国的工艺美术运动中，在德意志制造联盟的两个主要领导人赫尔曼·穆特修斯和亨利·凡·德·费尔德之间的冲突。穆特修斯也是一位德国政府设计教育界的领导人，他是标准化的拥护者。他信奉在实惠的装饰艺术前提下的大规模生产。亨利·凡·德·费尔德则认为大规模生产是对创造性和个性的威胁。

虽然设计师自发的个性比设计的历史风格变得更加重要，某些倾向脱颖而出：改革新哥特式影响、乡村和村舍式的外观、重复设计、纵向和拉长形式。为了表达出手工艺品固有的美，一些产品故意留下少许未完成的地方，造成一定质朴又富有活力的效果。另外，这项运动还有社会主义意味，明确地说，主要是在英国，另一个主要目的是手工艺者从他们做的东西中获得了满足。这一运动的支持者认为，这种满意完全否定了分工细致的机械产品的工业化进程。

事实上，工艺美术运动的支持者反对劳动分工原则，某些情况下机器的参与或者不参与是互不相关的。他们赞成工艺大师的想法，例如，在一些学徒的帮助下，创造一个家具项目的所有部分，同时也参与到它的装配和整理。这和一些工作环境是相反的，例如，在法国的制造厂，一切都服务于尽可能最快的生产。（例如，一个人或者一个团队处理一件家具所有的腿，一些人处理所有的面板，一些人组装零部件，还有另一些人描画和涂漆或者处理其他整理工作，均按照家具设计师制定的计划，但是设计师自己在其项目的创作中从来不参与实际工作。）工艺美术运动设法重新组合那些人类工作，使设计师亲自动手参与创造的每一步。该运动的一些最著名的倡导者，如莫里斯，更愿意为机械生产设计产品，这并不涉及可悲的分工和工艺人才的损失，但是他们却谴责这一点。莫里斯设计了无数机器生产的系列地毯。

2. 对今后艺术的影响

（1）欧洲

在欧洲广泛开展的工艺美术运动所体现的否定历史主义的质朴而诚实使用材料启发了设计师，如亨利·凡·德·费尔德和一些运动。如新艺术派、荷兰风格派运动、维也纳分离派，最终产生了包豪斯。该运动可称为现代主义的前奏，现代主义是纯粹的形式，被剥去了历史联系，终将回归到工业生产。

在俄罗斯，维克多·哈尔德曼，维克多·瓦斯涅佐夫和其他与阿布拉姆采沃殖民地相关的艺术家寻求在繁荣英国这项独立运动中振兴中世纪俄罗斯装饰艺术的精神和品质。

维也纳工场于1903年由约瑟夫·霍夫曼和科罗曼·莫泽创立，它的"维也纳工厂风格"在现代主义发展中独具一格。

英国在第二次世界大战期间的实用家具设计风格简单并以工艺美术的概念为基础。

坐落在爱尔兰科克市科克大学的霍南教堂建成于1916年，它是国际公认的爱尔兰工艺美术运动的代表。

（2）美国

在美国，工艺美术运动呈现出特殊的资产阶级风情。当欧洲运动试图重新建立被工业化破坏了的手工艺劳动的美好世界时，美国试图建立一种新型模式来代替个人英雄主义式的工艺生产：有品位的中产阶级家庭。他们认为艺术和手工装饰艺术中简单而精致的美学将使工业消费主义的新型体验变得高尚，使个人更理性，社会更和谐。简而言之，美国工艺美术运动是审美对应当代政治运动：进步主义。

在美国，工艺美术运动产生了各种各样为美国重新解读欧洲工艺美术思想的尝试。这些包括工匠风格的建筑、家具和其他装饰艺术，如古斯塔夫·斯蒂克利在他的杂志《艺术家》中所提倡的设计。大量对斯蒂克利家具（常被归类为"使命派风格"设计）的模仿，包括由他的兄弟成立的三个公司——埃伯特·哈伯德成立的伊卡洛夫托社区，弗兰克·劳埃德·赖特的草原学校（乡村学校运动），被格林兄弟推广普及的平房式住宅，像Byrdcliffe与玫瑰谷那样的乌托邦式社区，发展了由赫伯特·J. 哈普古德（Herbert J. Hapgood）建造的以平房群和庄园为特点的建筑，如山区湖新泽西，还发展了当代的工作室工艺运动。陶艺工作室，如"Grueby"，"Newcomb"，"Teco"，"Overbeck"和"Rookwood"陶器，英国的伯纳德·利奇，玛丽·佩里·斯特拉在底特律的弗瑞尔，还有欧内斯特·A. 巴彻尔德在加州帕萨迪纳的艺术瓷砖，查理·鲁尔夫画廊独特的家具也显示了工艺美术运动的明显影响。使命派风格、草原风格以及加州别墅风格的住宅在当今的美国依然很流行。

Free Reading 1

The Great Exhibition

Great Exhibition, 1851. Master-minded by Albert, the Great Exhibition was the largest trade show the world had ever seen. Joseph Paxton's Crystal Palace, spanning 19 acres within Hyde Park (London), was accepted after 233 other plans had been rejected. Some 6 million people between 1 May and 11 October 1851, many of them on railway excursions, visited 100,000 exhibits. Queen Victoria, always keen on her husband's achievements, visited 34 times. Profits secured land in Kensington, future sites for the Victoria and Albert Museum, the Science Museum, and the Natural History Museum.

The Great Exhibition of the Works of Industry of all Nations had its generic roots in a series of French National Exhibitions that had begun in Paris in 1798 where manufacturers from many branches of industry showed a wide range of products including ceramics, glass, furniture, and textiles. Just as the French National Exhibitions had been intended to restore French manufacturing industry to its former position of dominance in the wake of the political upheavals of the Revolution, so the Great Exhibition of 1851 was the culmination of a number of initiatives in the 1830s and 1840s to re-establish the position of British industry as the "workshop of the world" after a period of decline following the Napoleonic Wars. These included the establishment in 1935 of a Parliamentary Select Committee on The Arts and their Connection with Manufacturers and the subsequent institution of a national framework for design education. The enormous "Crystal Palace" went from plans to grand opening in just nine months, as shown in Figure 1.2 and Figure 1.3.

Figure 1.2 the Great Exhibition in the Crystal Palace

Figure 1.3 Exhibition interior

By 1849, the 11th French National Exhibition attracted 4,500 exhibitors and its scale and ambition resulted in key British design propagandists Henry Cole and Digby Wyatt being asked to report back on it to the *Royal Society of Arts* (RSA). In Britain, during the 1840s, the Royal Society of Arts had itself mounted a series of small-scale competitions promoting British industrial products that embraced the principles of artistic design. Such initiatives had brought Cole into the Society and, from 1847, developed into a series of annual exhibitions of industrial products culminating in the show of 1849, which attracted 73,000 visitors over a period of seven weeks. With the support of Prince Albert, the president of the RSA, and spurred on by Cole and Wyatt's 1849 report that the French were themselves considering an international exhibition, British ambitions were raised to do the same and a Royal Commission was swiftly established early in 1850 to oversee its development.

Although the RSA severed its formal connections with the exhibition as a result, a number of its key members continued to serve on the commission. At the Great Exhibition itself the public could admire many works that could be seen to embrace the interlinked fields of art, science, and manufacture, typified by a range of high-quality products manufactured by the Coalbrookdale Company. Amongst them were the ornamental gates to the exhibition, an Iron Dome, chairs, sculptures, and a variety of detailed ornamental castings, a number of which were designed by artists that had been employed by Henry Cole for his Felix Summer ley's art manufactures.

Amongst the many functional exhibits on display in the Crystal Palace the greatest impression was made by those from the United States of America that capitalized on the exploitation of standardization as a means of harnessing the true potential of mass-production technologies for mass markets. This outlook became known as the American System of Manufactures and was seen in products such as Colt's firearms, Hobbs' locks, McCormick's reaper and sewing machines. Its economic potential was sufficient to bring about the establishment of a Royal Commission that reported on the Machinery of the United States in 1854.

The Great Exhibition proved to be a highly profitable venture with a profit of £186,437, the surplus being used for educational purposes including the purchase of 87 acres (35 hectares) in South Kensington, London, as a centre for the arts and sciences. It was here that the 1862 International and the 1886 Colonial and Indian Exhibitions were later mounted and now housed a number of key buildings connected with the arts and sciences including the Victoria and Albert Museum, the Royal College of Art, Imperial College, and the Science Museum. The 1851 exhibition also stimulated a whole series of other international exhibitions, commencing with the Great Industrial Exhibition in Dublin and the World's Fair in New York in 1853, followed by the Exposition Universelle in Paris in 1855.

Admission fees

Admission prices to the Crystal Palace varied according to the date of visitation, with ticket prices decreasing as the parliamentary season drew to an end and London traditionally emptied of wealthy individuals. Prices varied from 3 guineas per day, £1 per day, five shillings per day, down to one shilling per day. The one shilling ticket proved most successful amongst the industrial classes, with four and a half million shillings being taken from attendees in this manner.

Free Reading 2

William Morris

William Morris (1834-1896), one of the most versatile and influential men of his age, was the last of the major English romantics and a leading champion and promoter of revolutionary ideas as poet, critic, artist, designer, manufacturer, and socialist (shown in Figure 1.4).

William Morris (24 March 1834—3 October 1896) was an English architect, furniture and textile designer, artist, writer, and socialist associated with the Pre-Raphaelite Brotherhood and the English Arts and Crafts Movement.

Born in Walthamstow in East London, Morris was educated at Marlborough and Exeter College, Oxford. In 1856, he became an apprentice to Gothic revival architect G. E. Street. That same year he founded the Oxford and Cambridge Magazine, an outlet for his poetry and a forum for development of his theories of hand-craftsmanship in the decorative arts. In 1861, Morris founded a design firm in partnership with the artist Edward Burne-Jones, and the poet and artist Dante Gabriel Rossetti which had a profound impact on the decoration of churches and houses into the early 20th century. Morris's chief contribution was as a designer of repeating patterns for wallpapers and textiles, many based on a close observation of na-

Figure 1.4 William Morris by George Frederic Watts, 1870

ture. He was also a major contributor to the resurgence of traditional textile arts and methods of production.

Morris wrote and published poetry, fiction, and translations of ancient and medieval texts throughout his life. His best-known works include The Defence of Guenevere and Other Poems (1858), The Earthly Paradise (1868—1870), A Dream of John Ball and the utopian News from Nowhere.

Morris was an important figure in the emergence of socialism in Great Britain, founding the Socialist League in 1884. He devoted much of the rest of his life to the Kelmscott Press, which he founded in 1891. The 1896 Kelmscott edition of the Works of Geoffrey Chaucer is considered a masterpiece of book design.

Red House, Bexleyheath

For several years after his marriage Morris was absorbed in two intimately connected occupations: the building and decoration of a house for himself and Jane. Meanwhile he was slowly abandoning painting; none of his paintings are dated later than 1862.

Red House at Bexleyheath in Kent (shown in Figure 1.5), so named when the use of red brick without stucco was a startling novelty in domestic architecture, was built by Phillip Webb to designs by Webb and Morris. It was Webb's first building as an independent architect and the first serious attempt made in Victorian England to apply art throughout to the practical objects of common life. Red House featured ceiling paintings by Morris, wall-hangings designed by Morris and worked by himself and Jane; furniture painted by Morris and Rossetti, and wall-paintings and stained-and painted glass designed by Burne-Jones.

Figure 1.5 Red House

Historic Preservation

Although Morris never became a practising architect, his interest in architecture continued throughout his life. In 1877 he founded the Society for the Protection of Ancient

Buildings (sometimes known as "Anti-Scrape"). His preservation work resulted indirectly in the founding of the National Trust. Combined with the inspiration of John Ruskin—in particular his essay "The Nature of Gothic" —architecture played an important symbolic part in Morris's approach to socialism.

Core Text 3

Art Nouveau

Art Nouveau is an international movement and style of art, architecture and applied art—especially the decorative arts—that peaked in popularity at the turn of the 20th century (1890 – 1905). The name "Art Nouveau" is French for "new art". It is also known as Jugendstil, German for "youth style", named after the magazine *Jugend* (shown in Figure 1.6), which promoted it, and in Italy, Stile Liberty from the department store in London, Liberty & Co., which popularized the style. A reaction to academic art of the 19th century, it is characterized by organic, especially floral and other plant-inspired motifs, as well as highly-stylized, flowing curvilinear forms. Art Nouveau is an approach to design according to which artists should work on everything from architecture to furniture, making art part of everyday life.

Figure 1.6 This front cover of an 1896 edition of the German magazine *Jugend* is decorated in Art Nouveau motifs. *Jugend* was strongly associated with the style and the magazine's name inspired the German term for the movement, *Jugendstil* ("*Jugend*" -style).

International Expos

A high point in the evolution of Art Nouveau was the *Exposition Universelle* of 1900 in Paris, which presented an overview of the "modern style" in every medium. It achieved further recognition at the *Esposizione Internazionale d'Arte Decorativa Moderna* of 1902 in Turin, Italy, where designers exhibited from almost every European country where Art Nouveau was practiced.

Belgium, Switzerland and France

In Paris, France, Maison de l'Art Nouveau, at the time run by Siegfried Bing. Artists such as Louis Majorelle and Victor Prouvé in Nancy, France, founded the Ecole de Nancy, giving Art Nouveau a new influence. In Brussels, Belgium the style was actively developed with the help of Victor Horta and Henry Van de Velde. Other Art Nouveau designers in Belgium, Switzerland and France include Theophile Alexandre Steinlen, Alphonse Mucha, Hector Guimard and Émile Gallé.

Germany

German Art Nouveau is commonly known by its German name, Jugendstil. Drawing from traditional German printmaking, the style uses precise and hard edges, an element that was rather different from the naturalistic style of the time. The movement was centered in Hamburg and was an essential element of the German movement. Within the field of Jugendstil art, there is a variety of different methods, applied by the various individual artists. Methods range from classic to romantic.

Britain

In the United Kingdom, Art Nouveau developed out of the Arts and Crafts Movement. The first stirrings of an Art Nouveau "movement" can be recognized in the 1880s, in a handful of progressive designs such as the architect-designer Arthur Mackmurdo's book cover design for his essay, published in 1883. Some free-flowing wrought iron from the 1880s could also be adduced, or some flat floral textile designs, most of which owed some impetus to patterns of High Victorian design. The most important center in Britain eventually became Glasgow, with the creations of Charles Rennie Mackintosh and his circle.

Key Words

[1] Art Nouveau [ɑːt, ˈnuːvəʊ] 新艺术主义；新艺术运动

[2] Jugendstil [juːgənt_ʃtiːl] [德语] 青春风格

[3] organic [ɔːˈɡænɪk] adj. 1. 器官的；器质性的 2. 有机(体)的，有机物的 3. 有机的；不使用化肥的；绿色的 4. 有机的；统一的；关联的 5. 逐渐的；演进的；自然的

[4] floral [ˈflɔːrəl] adj. 1. 用花做的 2. 用花装饰的

[5] curvilinear forms 曲线形式

[6] printmaking [ˈprɪnt, meɪkɪŋ] n. 版画复制(术)

[7] Arts and Crafts 工艺美术运动

[8] Ecole de Nancy 南锡艺术流派

Key Sentences

1. Art Nouveau is an international movement and style of art, architecture and applied art—especially the decorative arts—that peaked in popularity at the turn of the 20th century (1890—1905).

新艺术运动是一项国际运动和一种艺术风格、建筑和应用艺术—特别是装饰艺术—它在19世纪末，20世纪初达到流行顶峰。

2. A reaction to academic art of the 19th century, it is characterized by organic, especially floral and other plant-inspired motifs, as well as highly-stylized, flowing curvilinear forms. Art Nouveau is an approach to design according to which artists should work on everything from architecture to furniture, making art part of everyday life.

新艺术运动在19世纪的学术艺术中也产生了一定的反响，它的特点是有机的，特别是用花和其他植物作为装饰图案的灵感，具有高度的风格化和流动的曲线形式。新艺术运动是一种设计方法，艺术家将其应用到从建筑到家具等日常生活中的每一部分。

课文翻译

新艺术运动

新艺术运动是一项国际运动和一种艺术风格、建筑和应用艺术——特别是装饰艺术——在19世纪末，20世纪初达到流行顶峰。"Art Nouveau"的名称来自法语，意为"新艺术"。也被称为"Jugendstil"，德语意为"青春风格"，得名于《青春》杂志。在意大利，"Stile Liberty"来自于伦敦的百货公司Liberty商店，该公司强化了风格的流行。新艺术运动在19世纪的学院艺术中也产生了一定的反响，它的特点是有机的，特别是用花和其他植物作为装饰图案的灵感，具有高度的风格化和流动的曲线形式。新艺术运动是一种设计方法，艺术家将其应用到从建筑到家具等日常生活中的每一部分。

国际博览会

新艺术运动的一个发展高潮是1900年在巴黎举办的世界博览会，其中，在每一种媒介中都展现了对"现代风格"的综述。在1902年意大利都灵举办的国际现代装饰艺术展览上，新艺术获得了进一步的认知。在那里，参展的设计师来自几乎所有进行了新艺术实践的欧洲国家。

比利时、瑞士和法国

位于法国巴黎的新艺术之屋，当时由萨姆尔·宾开办。法国南锡的艺术家路易斯·若雷勒和维克多·普鲁韦建立了南锡艺术流派，对新艺术运动产生了新的影响。在比利时的布鲁塞尔，这种风格在维克多·霍尔塔和亨利·凡·德·威尔德的帮助下得到积极发展。其他在比利时、瑞士、法国的新艺术设计师包括泰奥菲勒·亚历山大·斯坦伦、阿方·木栅、赫克托·吉马德和艾米里加利。

德国

德国新艺术运动以其德语名称"青春风格"而闻名。传统的德国版画风格精细，边缘尖硬，这与当时的自然主义风格有所不同。这场运动以汉堡为中心，是德国运动中的一个非常重要的元素。在青春风格艺术的领域中，多种不同风格的独立艺术家运用了多种不同的方法。这些方法的范围从经典到浪漫。

英国

在英国，新艺术运动起源于工艺美术运动。新艺术运动的第一次出现被认定在19世纪80年代的一些先进的设计中，例如，建筑设计师阿瑟·马克穆多为他的论文所做的书籍封面设计，该书出版于1883年。19世纪80年代自由风格的铸铁设计也是例证之一，或者一些植物装饰的纺织图案设计，其中大部分要归功于高级维多利亚模式的促进。查尔斯·伦尼·麦金托什和他的团队的创作最终使格拉斯哥成为英国最重要的中心。

Free Reading 1

Art Nouveau in France

Paris and Nancy

The new style became most firmly established in France where it retained its popularity the longest. As Rococo and Neo-Rococo achieved their culmination in France, it is only natural that the most importantinfluence of Rococo on Art Nouveau occurred in this cultural area.

Figure 1.7　Emile Gallé

Two distinct centers developed in France, one in Paris around S. Bing and the other in Nancy under the aegis of Emile Gallé (shown in Figure 1.7). The closest ties between Rococo and the new style occurred in Nancy, where in the 18th century the earliest Neo-Rococo furniture was made and flourished until it was supplanted by Napoleon III's Neo-Classicism in the 1860's. The artistic revival of the 1890's once again brought Rococo into the foreground, as Nancy's artists blended it with the new art.

Emile Gallé

The undeniably important position held by France in Art Nouveau was chiefly won by Gallé and a host of other artists and designers. Gallé, whose reputation to a large extent is based on hiswork in glass, was the first great exponent of this style in France. He was one of the most interesting of all designers, as well as France's outstanding naturalist.

By around 1900 Gallé's workshop employed approximately three hundred workers, who were busy making glass and furniture. To a certain extent Gallé's designs for furniture were in the French stylistic traditions, such as Louis XIV, Louis XV or Louis XVI. In a word, the construction and design were traditional. Only the decoration, carving which he developed two-dimensionally andmarquetry which was Gallé's specialty and varied from plant forms to verses, displayed dramatic tendencies.

Gallé believed that the function of furniture should find expression through decoration and not construction. Gallé's inscriptions gave a symbolical expression to the idea of his furniture; for example, a work table inscribed "Travail est Joie" (Work is Joy). This class of furniture bearing inscriptions, which induced an aesthetic experience beyond what was innate in the piece of furniture itself, was much in fashion in France, where it was called "meubles parlants" (furniture that talks).

Louis Majorelle

None of Gallé's furniture ever achieved the daring unconventionality of his contemporary Louis Majorelle, who may be assigned a position among Art Nouveau furniture designers which corresponds to that of Gallé in art glass. Shortly after Majorelle inherited his father's furniture business in 1879, he began to design furniture in the fashionable Louis XIV, Louis XV and Louis XVI styles. However, he worked mainly in the Louis XV style or Neo-Rococo, until around the end of the century, when, under Gallé's influence, he started to design in the Art Nouveau taste.

Less bound by tradition in construction and more plastic in conception than Gallé, the main characteristic of his furniture is the graceful but powerful dynamic line. The plastic conception was so important to him that he worked in clay as a sculptor and molded his most important furniture forms, later translating the model into carved wood. No doubt his finest work ranks among the most perfected achievements of the Art Nouveau style. The representative works are shown in Figure 1. 8.

Figure 1.8　Lit "Nénuphars" Louis Majorelle

Emile André

Unlike Gallé and Majorelle the other Nancy furniture designers were architects. Among the outstanding members were Emile André who was Nancy's most important architect, Jacques Gruber and Eugène Vallin (1865—1925).

The furniture designs of the latter are more in the manner of Majorelle than any of the designers belonging to the Nancy school. Emile André adopted a similar sculptural and dynamic form language based on floral and Neo-Baroque inspiration.

Later, in the first few years of the 1900's, he, like Majorelle, discarded the floral elements in favor of a more abstract decoration or no decoration at all, relying on the graceful flow of the structure to provide the decorative accent or element. This style with large and smooth surfaces was simple, sober and more international, and lasted to a certain extent until just about the beginning of the First World War. After Gallé's death in 1904 the Nancy school, devoid of his leadership and inspiration, faded from the limelight, as Nancy returned once again to the province's own enduring style, Louis XV.

Siegfried Bing

In contrast to the Nancy school, Parisian furniture is less ponderous, more refined and restrained. The Nature-inspired decoration is more stylized, at times abstract and often restricted to small areas. Unlike the Nancy school, where the artists followed in Gallé's footsteps, the Parisian designers were completely individual personalities, each presenting his own form of the new style.

S. Bing, a native of Hamburg, art dealer and critic, opened his first shop, Maison de l'Art Nouveau, at 22 Rue de Provence, in December 1895. This was to become the focal point in Paris for the new style. Apart from the three most prominent who worked for Bing, Georges de Feure (1869—1928), Eugène Gaillard and Eugène Colonna, there were a number of other well-known representatives, such as Alexandre Charpentier (1856—1909) and Felix Aubert (b. 1866), Tony Selmersheim (b. 1871), and Hector Guimard (1867—1942). This group produced some of the finest examples of Art Nouveau furniture.

Figure 1.9 Metro paris

No doubt the leading architect was Guimard. The symbolical plant conception that clearly influenced the construction of his famed Metro stations (See Figure 1.9) in Paris is also evident in his furniture. Art Nouveau was frequently dubbed Style Metro by the general public. Of all Bing's artists Georges de Feure appears the most conservative, for although his furniture is certainly Art Nouveau it assumes a traditional French character. Until 1900 the form language he uses for his constructive elements is always derived from plants and flowers with stalks, and is notable for its elegant and delicate execution.

In contrast to de Feure, the work of Eugène Gaillard is more virile; at times ponderous. His plastics and dynamic approach conveys a quality of forceful and powerful movement that relates to Majorelle. One of his important and striking pieces at the Paris Exhibition of 1900 was a tall cupboard with completely abstract and plastic ornament.

As a furniture designer Eugène Colonna may be placed midway between the two. He can be either elegant or dynamic, as well as being more austere, more severe in his decoration than either one of them.

End of Art Nouveau in France

After the Paris Exhibition of 1900 when the Parisian artists were faced with the exaggerations of the style, they retreated and made a subtle return to a quiet, restrained form of Art Nouveau, which with its inventive and graceful elegance was more in accord ance with the true traditions of French period furniture. It was a return to a simplified and modernized Classicism, especially in the direction of Directoire and Empire.

Free Reading 2

Antoni Gaudí

Antoni Gaudí was born in the province of Tarragona in southern Catalonia on 25 June 1852. The artist's parents, Francesc Gaudí Serra and Antònia Cornet Bertran, both came from families of coppersmiths.

During his youth, Antoni Gaudi suffered many times from the rheumatic fevers that were common at the time. This illness caused him to spend much time in isolation, and it also allowed him to spend lots of time alone with nature. It was this exposure to nature at an early age which is thought to have inspired him to incorporate natural shapes and themes into his later work.

Gaudí was a devout Catholic, to the point that in his later years he abandoned secular work and devoted his life to Catholicism and his Sagrada Família. He designed it to have 18 towers, 12 for the 12 apostles, 4 for the 4 evangelists, one for Mary and one for Jesus. Soon after, his closest family and friends began to die. His works slowed to a halt, and his attitude changed. One of his closest family members—his niece Rosa Egea—died in 1912, only to be followed by a "faithful collaborator", Francesc Berenguer Mestres, two years later. After these tragedies, Barcelona fell on hard times economically. The construction of La Sagrada Família slowed; the construction of La Colonia Güell ceased altogether. Four years later in 1916, Eusebi Güell, his patron, died.

Perhaps it was because of this unfortunate sequence of events that Gaudí changed. He became reluctant to talk with reporters or have his picture taken and solely concentrated on his masterpiece, La Sagrada Família. He spent the last few years of his life living in the crypt of the "Sagrada Familia".

On 7 June 1926, Gaudí was run over by a tram. Because of his ragged attire and empty pockets, many cab drivers refused to pick him up for fear that he would be unable to pay the fare. He was eventually taken to a paupers' hospital in Barcelona. Nobody recognized the injured artist until his friends found him the next day. When they tried to move him into a nicer hospital, Gaudí refused, reportedly saying "I belong here among the poor." He died three days later on 10 June 1926, at age 73, half of Barcelona mourning his death. He was buried in the midst of La Sagrada Família.

Although Gaudí was constantly changing his mind and recreating his blueprints, the only existing copy of his last recorded blue prints was destroyed by the anarchists in 1938 during the Spanish Civil War. This has made it very difficult for his workers to complete the cathedral in the fashion Gaudí most likely would have wished. It is for this that Gaudí is known to many as "God's Architect". La Sagrada Família is now being completed, but differences between his work and the new additions can be seen. As of 2007, completion of the Sagrada Família is planned for 2026. They wish to do this because it is the anniversary of his death.

It is widely acknowledged that Gaudí is a part of Barcelona. His first works were designed in the style of gothic architecture and traditional Catalan architectural modes, but he soon developed his own distinct sculptural style. French architect Eugene Viollet-le-Duc, who promoted an evolved form of gothic architecture, proved a major influence on Gaudí. Some of his greatest works, most notably La Sagrada Família, have an almost hallucinatory power.

Gaudí, throughout his life, studied nature's angles and curves and incorporated them into his designs and mosaics. Instead of relying on geometric shapes, he mimicked the way men stand upright. The hyperboloids and paraboloids he borrowed from nature were easily reinforced by steel rods and allowed his designs to resemble elements from the environment.

Because of his rheumatism, the artist observed a strict vegetarian diet, used homeopathic drug therapy, underwent water therapy, and hiked regularly. Long walks, besides suppressing his rheumatism, further allowed him to experience nature. Gaudí loved for his work to be created by nature as he used concrete leaves and vine windows to create his ideas for him, so his work is not just because of him but because of nature as well.

Gaudí's originality was at first ridiculed by his peers. As time passed, though, his work became more famous. He stands as one of history's most original architects. Some works of Gaudí are as follows: Figure 1.10: The Casa Milà, Barcelona, Figure 1.11: Gaudí's unfinished masterpiece, Sagrada Família, currently under construction

Figure 1.10　The Casa Milà, Barcelona

Figure 1.11　Gaudí's unfinished masterpiece, Sagrada Família, currently under construction

Free Reading 3

Art Deco

Movement in design, interior decoration, and architecture in the 1920s and 1930s in Europe and the U.S, The name derives from the Exposition Internationale des Arts Décoratifs et Industriels Modernes in Paris in 1925. Its products included both individually crafted luxury items and mass-produced wares, but, in either case, the intention was to create a sleek and anti-traditional elegance that symbolized wealth and sophistication. Influenced by Art Nouveau, Bauhaus, Cubist, Native American, and Egyptian sources, the distinguishing features of the style are simple, clean shapes, often with a "streamlined" look; ornament that is geometric or stylized from representational forms; and unusually varied, often expensive materials, which frequently include man-made substances (plastics, especially bakelite; vita-glass; and ferroconcrete) in addition to natural ones (jade, silver, ivory, obsidian, chrome, and rock crystal). Typical motifs included stylized animals, foliage, nude female figures, and sun rays. New York City's Rockefeller Center (especially its interiors supervised by Donald Deskey), the Chrysler Building by William

Van Alen, and the Empire State Building by Shreve, Lamb & Harmon are the most monumental embodiments of Art Deco.

The visual origins of the style included Cubist painting (particularly the more two-dimensional forms of Synthetic Cubism), the vivid colours associated with Matisse and the Fauves, the abstracted botanical and zoological forms explored by Raoul Dufy and members of Paul Poiret's Atelier Martine. Léon Bakst's striking and often orientalizing stage and costume designs for Diaghilev's Ballets Russes were also powerful ingredients. A widespread interest in ethnography, "primitivism", and the collecting of "primitive" artefacts in the early years of the 20th century also informed many designers' use of exotic woods, snakeskin, ivory, and other materials drawn from the French colonies.

The dissemination of French decorative art was aided by the launching of a number of French luxury liners such as the Paris (1921), the le de France (1927), and the Normandie (1935). Like major international exhibitions these floating palaces were symbols of national prestige. Often subsidized by the French government they afforded significant opportunities for French artistes-decorators to bring their work to an international, often wealthy, transatlantic traveling public. Important too in transmitting many of its features was the increasingly powerful and popular medium of film, especially the output of Hollywood, which often drew on Art Deco as a basis for its most striking sets. Pivotal in this were the highly glamorous sets overseen by art directors such as Cedric Gibbons of MGM (including Our Dancing Daughters, 1928, Our Modern Maidens, 1929, and Our Blushing Brides, 1930) and Van Nest Polglase of RKO (including the "latest idea in interior architecture for the modern home" of The Magnificent Flirt, 1928, Flying Down to Rio, 1933, and Top Hat, 1935).

Art Deco was also seen in Britain, often in the form of geometric sunburst motifs found on tea services, garden gates and fences, stained glass windows in domestic hallways, and radio cabinet loudspeaker grilles. In addition to enjoying Deco in the luxury film sets on the cinema screen, the general public also experienced it in the design of leisure architecture including hotels, theatres, lidos, and cinemas. Typical of the latter were the interiors and exteriors of the Odeon cinemas, characterized by the decorative manipulation of abstract forms, finishes, and colours. Some works of Art Deco as follows, Figure 1.12: "The Musician", oil painting on canvas by Tamara de Lempicka, 1929, Figure 1.13: Terracotta sunburst design in gold behind sky blue and deep blue above the front doors of the Eastern Columbia Building in Los Angeles.

Attributes

Art Deco was an opulent style, and its lavishness is attributed to reaction to the forced austerity imposed by World War I. Its rich, festive character fitted it for "modern" contexts, including the Golden Gate Bridge, interiors of cinema theaters (a prime example being the Paramount Theater in Oakland, California) and ocean liners such as the Île de

Figure 1.12 "The Musician", oil painting on canvas by Tamara de Lempicka, 1929

Figure 1.13 Terracotta sunburst design in gold behind sky blue and deep blue above the front doors of the Eastern Columbia Building in Los Angeles

France, the Queen Mary, and Normandie. Art Deco was employed extensively throughout the United States' train stations in the 1930s, designed to reflect the modernity and efficiency of the train. Art Deco made use of many distinctive styles, but one of the most significant of its features was its dependence upon a range of ornaments and motifs. The style is said to have reflected the tensions in the cultural politics of its day, with eclecticism having been one of its defining features. In the words of Scott Fitzgerald, the distinctive style of Art Deco was shaped by all the nervous energy stored up and expended in the War. Art Deco has been influenced in part by movements such as Cubism, Russian Constructivism and Italian Futurism, which are all evident in Art Deco decorative arts.

Materials and Design

Art Deco is characterized by use of materials such asaluminium, stainless steel, lacquer and inlaid wood. Exotic materials such as sharkskin (shagreen), and zebraskin were also in evidence. The bold use of stepped forms and sweeping curves (unlike the sinuous, natural curves of the Art Nouveau), chevron patterns, and the sunburst motif are typical of Art Deco. Some of these motifs were ubiquitous—for example, sunburst motifs were used in such varied contexts as ladies' shoes, radiator grilles, the auditorium of the Radio City Music Hall, and the spire of the Chrysler Building.

Unit Two
Modernism and Industrial Design

Core Text 4

Constructivism

Constructivism

Constructivism is a movement in modern art originating in Moscow in 1920 and characterized by the use of industrial materials such as glass, sheet metal, and plastic to create nonrepresentational, often geometric objects.

Constructivism is avant-garde tendency in the 20th-century painting, sculpture, photography, design and architecture, with associated developments in literature, theatre and film. The term was first coined by artists in Russia in early 1921 and achieved wide international currency in the 1920s. Russian Constructivism refers specifically to a group of artists who sought to move beyond the autonomous art object, extending the formal language of abstract art into practical design work. This development was prompted by the Utopian climate following the October Revolution of 1917, which led artists to seek to create a new visual environment, embodying the social needs and values of the new Communist order. The concept of International Constructivism defines a broader current in Western art, most vital from around 1922 until the end of the 1920s, that was centred primarily in Germany. International Constructivists were inspired by the Russian example, both artistically and politically. They continued, however, to work in the traditional artistic media of painting and sculpture, while also experimenting with film and photography and recognizing the potential of the new formal language for utilitarian design.

Art in the Service of the Revolution

As much as involving itself in designs for industry, the Constructivists worked on public festivals and street designs for the post-October revolution Bolshevik government. Perhaps the most famous of these was in Vitebsk, where Malevich's UNOVIS Group painted propaganda plaques and buildings (the best known being El Lissitzky's poster *Beat the Whites with the Red Wedge* (1919)), inspired by Vladimir Mayakovsky's declaration "the streets our brushes, the squares our palettes", artists and designers participated in public life throughout the Civil War. A striking instance was the proposed festival for the Comintern congress in 1921 by Alexander Vesnin and Liubov Popova, which resembled the constructions of the OBMOKhU exhibition as well as their work for the theatre.

As a part of the early Soviet youth movement, the constructivists took an artistic outlook aimed to encompass cognitive, material activity, and the whole of spirituality of mankind. The artists tried to create works that would take the viewer out of the traditional setting and make them an active viewer of the artwork. In this it had similarities with the Russian Formalists' theory of "making strange", and accordingly their leading theorist Viktor Shklovsky worked closely with the Constructivists, as did other formalists like Osip Brik. These theories were tested in the theatre, particularly in the work of Vsevolod Meyerhold, who had set up what he called "October in the theatre". Meyerhold developed a "biomechanical" acting style, which was influenced both by the circus and by the "scientific management" theories of Frederick Winslow Taylor. Meanwhile the stage sets by the likes of Vesnin, Popova and Stepanova tested out Constructivist spatial ideas in a public form.

Tatlin, "Construction Art" and Productivism

The canonical work of Constructivism was Vladimir Tatlin's proposal for the Monument to the Third International (1919) which combined a machine aesthetic with dynamic components celebrating technology such as searchlights and projection screens. Gabo publicly criticized Tatlin's design saying *Either create functional houses and bridges or create pure art, not both*. This had already led to a major split in the Moscow group in 1920 when Gabo and Pevsner's *Realistic Manifesto* asserted a spiritual core for the movement. This was opposed to the utilitarian and adaptable version of Constructivism held by Tatlin and Rodchenko. Tatlin's work was immediately hailed by artists in Germany as a revolution in art: a 1920 photo shows George Grosz and John Heartfield holding a placard saying "Art is Dead—Long Live Tatlin's Machine Art", while the designs for the tower were published in Bruno Taut's magazine *Fruhlicht*.

Constructivism and Consumerism

In 1921, a New Economic Policy was set in place in the Soviet Union, which reintroduced a limited state capitalism into the Soviet economy. Rodchenko, Stepanova, and others made advertising for the co-operatives that were now in competition with commercial businesses. The poet-artist Vladimir Mayakovsky and Rodchenko worked together and called themselves *"advertising constructors"*. Together they designed eye-catching images featuring bright colours, geometric shapes, and bold lettering. The lettering of most of these designs was intended to create a reaction, and function on emotional and substantive levels—most were designed for the state-run department store Mosselprom in Moscow, for pacifiers, cooking oil, beer and other quotidian products, with Mayakovsky claiming that his "nowhere else but Mosselprom" verse was one of the best he ever wrote.

In addition, several artists tried to work in clothes design with varying levels of success: Varvara Stepanova designed dresses with bright, geometric patterns that were mass-produced, although workers' overalls by Tatlin and Rodchenko never achieved this and remained prototypes. The painter and designer Lyubov Popova designed a kind of Constructivist flapper dress before her early death in 1924, the plans for which were published in the journal *LEF*. In these works Constructivists showed a willingness to involve themselves in fashion and the mass market. Some works of Constructivism are shown in Figure 2.1, 2.2, 2.3, 2.4.

Figure 2.1　Tatlin Tower, Model of the *Monument to the Third International*

Figure 2.2　Photograph of the first Constructivist Exhibition, 1921

Figure 2.3　Agitprop poster by Mayakovsky

Figure 2.4　Photomontage by Tatlin showing his clothing designs, 1924

Key Words

　　[1] constructivism [kənˈstrʌktɪvɪzəm] *n*. 构成主义，构成派

　　[2] avant-garde [ˌævɒŋˈgɑːd] *n*. 革新者，(尤指)艺术上的先锋派 *adj*. 新的，开拓的，先锋的；前卫的

　　[3] sculpture [ˈskʌlptʃə] *n*. 1. 雕刻，雕塑，雕像 2. 雕刻品 3. 雕刻术；雕塑术 *vt*.

& *vi.* 雕刻，雕塑

[4] utilitarian [juːˌtɪlɪˈteərɪən] *adj.* 1. 有效用的；实用的 2. 功利（主义）的 *n.* 功利主义者；实用主义者

[5] encompass [ɪnˈkʌmpəs] *vt.* 1. 围绕；包围 2. 包含，包括，涉及（大量事物）

[6] substantive [ˈsʌbstəntɪv] *adj.* 1. 真的，真实的，实际的 2. （军衔）永久的；非临时的 3. 实质性的；本质上的；重大的；严肃认真的 *n.* 真实

[7] flapper [ˈflæpə] *n.* （打苍蝇等的）拍子，蝇拍

[8] communist [ˈkɔmjunɪst] *n.* 共产主义者 *adj.* 共产主义的

Key Sentences

1. Russian Constructivism refers specifically to a group of artists who sought to move beyond the autonomous art object, extending the formal language of abstract art into practical design work. This development was prompted by the Utopian climate following the October Revolution of 1917, which led artists to seek to create a new visual environment, embodying the social needs and values of the new Communist order.

俄罗斯构成主义涉及一群特别的艺术家，他们尝试超越自主的艺术作品，将抽象的形象语言扩展到实际的设计工作中。这种发展在1917年十月革命后被乌托邦思潮推进。它引导艺术家努力创建一个新的视觉环境，表现了社会的需求和新共产主义秩序的价值。

2. As a part of the early Soviet youth movement, the constructivists took an artistic outlook aimed to encompass cognitive, material activity, and the whole of spirituality of mankind. The artists tried to create works that would take the viewer out of the traditional setting and make them an active viewer of the artwork.

作为早期的苏维埃青年运动的一部分，构成主义者持有一种艺术观点，着眼于围绕认知、物质活动以及人类的精神性。艺术家们试着去创造作品，使观众走出传统的背景，并且使他们成为艺术作品积极的欣赏者。

课文翻译

构 成 主 义

构成主义是1920年发生在莫斯科的现代艺术运动，其特征是利用工业材料，如玻璃、金属片和塑料等来创造抽象的几何形体物品。

构成主义是20世纪绘画、雕塑、摄影、设计和建筑的先锋，在文学、戏剧和电影领域也有相关发展。这个概念首次于1921年由俄罗斯艺术家提出，并在20世纪20年代成功地在国际上广泛流行。俄罗斯构成主义涉及一群特别的艺术家，他们尝试超越自主的艺术作品，将抽象的形象语言扩展到实际的设计工作中。这种发展在1917年十月革命后被乌托邦思潮推进，它引导艺术家努力去创建一个新的视觉环境，表现了社会的需求和新共产主义秩序的价值。国际构成主义的概念在西方艺术中定义了一个明显的趋势，从1922年左右最重要的时期到20世纪20年代末期，主要集中在德国。国际构成主义从俄罗斯的艺术和政治案例获取灵感。但是，他们的工作不仅仅是传统绘画和雕塑，同时也尝试着在电影、摄影以及为实用设计寻找潜在的，新的语言形式。

艺术在革命中的作用

为了尽可能多地将自身涉入设计行业，构成主义者为后十月革命布尔什维克政府工作，为其设计公共节日和街道。或许其中最著名的当数在维帖布斯克（白俄罗斯东北部城市），由马列维奇的"UNOVIS"团队绘制了宣传板和建筑（最著名的是埃尔·利西茨基 1919 年的海报"红色铁杆打击白色资本主义"），灵感来自于弗拉基米尔·马雅科夫斯基的宣言"街道是我们的画笔，广场是我们的调色板"。在内战期间，艺术家和设计师们参与到公共生活当中。最突出的例子就是亚历山大·维斯宁和柳波夫·波波娃为 1921 年的共产国际议会所提议的庆祝节目，它类似于"OBMOKhU"展览会的建筑，也像他们为之工作的剧院。

作为早期的苏维埃青年运动的一部分，构成主义者持有一种艺术观点，着眼于围绕认知、物质活动，以及人类的整体精神性。艺术家们试着去创造作品，使观众走出传统的背景，并且使他们成为艺术作品积极的欣赏者。这方面与俄罗斯形式主义的"制造奇特"的理论相似，并且，其主要的理论家维克托·什克洛夫斯基与构成主义者工作更为密切，与形式主义者奥斯普·布瑞克也很密切。这些理论在戏剧中有所尝试，特别是弗谢沃洛德·梅耶荷德的工作，他创立了"戏剧中的十月"。梅耶荷德发展了一种被称为"生态机械化"的表演形式，受到了马戏团和弗德雷克·维斯洛·泰勒的"科学化管理的理论"的影响。同时，舞台设计就好比维斯宁，波波娃和斯特帕诺娃所尝试的在公共形式中的结构主义空间概念。

塔特林，"建筑艺术"和生产主义

构成主义的权威作品是斐德拉米尔·塔特林在 1919 年为第三国际所做的纪念碑的提议，它将机械化审美和强有力的部件有机结合，如探照灯和投影屏。嘉博公然批评塔特林的设计，他认为要么建造功能性的房屋和桥梁，要么就是纯粹的艺术，而不是两者皆宜。1920 年，当嘉博和佩夫斯纳的现实主义宣言中断言运动精神核心的时候，这导致莫斯科团队发生了主要的分裂。这反对了塔特林和罗钦科所持的构成主义具有实用性和适应性的观点。塔特林的作品立即作为艺术革命在德国受到铺天盖地的欢迎：1920 年的照片显示，乔治·格罗希 和约翰·哈特菲尔德举着一块牌子，上面说，"艺术死了，塔特林的机械艺术长存"。同时，塔特林的设计在布鲁诺·陶特的杂志《Fruhlicht》发表。

构成主义和消费主义

1921 年，苏联适时颁布了一项新经济政策，政策重新引入了受限制的国家资本主义进入苏维埃经济。罗钦科，斯特帕诺娃和其他人为合作社制作广告，现在来说就是商业竞争。诗人艺术家弗拉基米尔·马雅科夫斯基和罗钦科一起合作，并称他们自己为"广告建设者"。他们共同设计了令人瞩目的图像，具有明亮的色彩、几何形状和粗体字。这些设计的目的是创建一种反应，并且作用于情感和实质层次——大部分是专为莫斯科国营百货商店 Mosselprom 做设计，包括奶嘴、食用油、啤酒及其他日常产品，马雅科夫斯基声称他的"惟独 Mosselprom"的表达是他所写过的最好的。

此外，一些艺术家在服装设计上取得了不同程度的成功。瓦尔瓦拉·斯特帕诺娃设计的明快的，带有几何图案的礼服被大规模生产，然而，塔特林和罗钦科所设计的工作服并没有取得这种成绩，只是留下了蓝本。油画家和设计师柳波夫·波波娃在 1924 年去世前设计了一款构成主义淑女服饰，刊登在《LEF》杂志上。在这些作品中，构成主义者显示，他们表达了投身到服装设计和大众市场中去的意愿，在那里，他们试图去平衡他们的共产主义信仰。

Free Reading

Piet Mondria

Piet Mondrian was born on March 7, 1872, in Amersfoort. His father, a schoolteacher, wished Piet to become a teacher, and he earned his diploma for teaching. But in 1892 he entered the Academy of Fine Arts in Amsterdam, where he studied for several years and was encouraged by artists of the Hague school. Mondrian's early pictures are mostly of such subjects as meadows with farms and cows or windmills. Although a few of his works from about 1900 show some influence of Claude Monet and symbolism, he continued working in a very conservative tradition for a number of years.

Development of His Style

In 1908 Mondrian became deeply involved in the latest developments in art, and in the course of the next 10 years or so he developed with astonishing rapidity through a succession of styles. He began to use pure, glowing colors and expressive brushwork under the influence of pointillism and Fauvism in pictures which are almost like those of Vincent Van Gogh in their vivid colors and intensity of expression.

By the time Mondrian moved to Paris in 1912, he had already seen a few cubist pictures and had begun to be influenced by cubism. But at a time when Georges Braque and Pablo Picasso were turning back to figuration, Mondrian decided to carry cubism through to what seemed to him its logical culmination of pure abstraction. Although he continued until as late as 1916 to make some reference to such subjects as trees and the facades of buildings, he gradually eliminated all traces of figuration. He quickly assimilated the cubist idiom of Braque and Picasso, working in grays or ochers and sometimes using an oval composition, but over the following years his compositions became more and more clarified, with a concentration on vertical and horizontal lines.

This development became particularly marked after Mondrian returned in 1914 to Holland, where, because of the outbreak of war, he remained until 1919, living mainly in the artists' colony at Laren. It can be seen, for instance, in various paintings and drawings of the sea in which the movement of waves is evoked by short horizontal and vertical lines (his so-called plus-and-minus compositions).

Advent of Neoplasticism

Only after Mondrian's return to Paris in 1919 did this tendency reach its culmination in the style to which he gave the name neoplasticism. From 1922 on he worked exclusively with vertical and horizontal lines and with white, black, and the three primary colors—the strongest and purest possible contrasts. In all but a few of his last works, he divided his pictures asymmetrically by a grid of heavy black vertical and horizontal lines, with certain rectangles painted a uniform intense red, blue, or yellow and all the other areas left a brilliant white. But within these limitations he achieved a wide range of effects by varying the proportions, the choice and distribution of the colors, and so on. Although he painted some pictures on canvases of square format hung diagonally, he always kept the lines strictly vertical and horizontal and indeed resigned from de Stijl in 1925 because Van Doesburg had introduced diagonal lines.

Mondrian lived in Paris from 1919 to 1938 and in London from 1938 to 1940; then he settled in New York. In his last works he used colored lines instead of black ones and even broke up the lines into a lively mosaic of different colors. Some his last works are shown in Figure 2.5, Figure 2.6. He died in New York City on Feb. 1, 1944.

Figure 2.5　Piet Mondrian, Composition II in Red, Blue, and Yellow, 1930

Figure 2.6　Piet Mondrian, Composition with Red, Yellow, and Blue 1937—1942

Core Text 5

Modernism

Modernism, also known as the Modern Movement, marked a conscious break with the past and has been one of the dominant expressions of design practice, production, and theory in the 20th century and is generally characterized visually by the use of modern materials such as tubular steel and glass, the manipulation of abstract forms, space and light, and a restrained palette, dominated by white, off-white, grey, and black. Following on from the well-known phrase "Ornament and Crime" coined by Adolf Loos as the title of an article of 1908, later echoed by Le Corbusier in his assertion that "trash is always abundantly decorated" was the notion that the surfaces were generally plain. When decoration was used it was restrained and attuned to the abstract aesthetic principles of the artistic avant-garde such as those associated with De Stijl or Constructivism. Also closely associated with Modernism was the maxim "form follows function" although in reality this was often more symbolic than the case in reality, a visual metaphor for the *Zeitgeist* or spirit of the age. Nonetheless Modernism found forms of material expression alongside exciting, new, and rapidly evolving forms of transport and communication, fresh modes of production and materials coupled to technological and scientific change, alongside a contemporary lifestyle powered by electricity.

The roots of Modernism lie in the design reform movement of the 19th century and were nurtured in Germany in the years leading up to the First World War. The Modernist legacy is considerable in terms of design (whether furniture, tableware, textiles, lighting, advertising, and typography or other everyday things), architecture (whether public or private housing, cinemas, office blocks, and corporate headquarters), or writings (theories, manifestos, books, periodicals, and criticism). This has done much to cement Modernism firmly into the mainstream history of design. Furthermore, it is also heavily represented in numerous museums around the world that have centred their design collections drawn from the later 19th century through to the last quarter of the 20th century around the Modernist aesthetic and its immediate antecedents.

After the Second World War the International Style was taken up by many major multinational companies for the architecture, interiors, furniture, and furnishings and equipment of their offices and showrooms, thus promoting themselves through their emphatically modern identity as efficient, up-to-date, and internationally significant organizations in a global economy. In the eyes of some, Modernism's earlier associations with social democratic ideals had been transmuted in its later manifestations to support capitalist ends. Used widely in design and architecture in the 1950s and 1960s, such stylistic traits also attracted increasing criticism from a younger generation of designers, architects.

Such trends found expression in the increasingly rich and vibrant vocabulary of Post-modernism, echoed in the increasingly ephemeral lifestyle enjoyed by those in the industrial world with greater levels of disposable income. Ideas about what was called "Good Design" in the 1950s and 1960s were formally linked to the Modernist aesthetic but without the social utopian underpinning promoted by many of the first generation of Modernists in the interwar years. In Britain such objects were approved by the state-funded Council of Industrial Design and seen in opposition to the elaborate styling and obsolescence inherent in American design that was becoming attractive to British consumers, whilst in the United States, at the Museum of Modern Art in New York, they were also seen as exemplars of European restraint.

A key text that has played an important role in defining Modernism has been Nikolaus Pevsner's widely read book, first published as *Pioneers of the Modern Movement* (1936). It has subsequently undergone substantial revisions (including a major one supported by the Museum of Modern Art, New York, in 1949) and numerous reprints under the title of *Pioneers of Modern Design: from William Morris to Walter Gropius*. Pevsner provides an account of the ways in which John Ruskin, William Morris, and exponents of the Arts and Crafts Movement fought against what they saw as the morally decadent and materialistic indulgence in historical ornamentation, inappropriate use of materials, and "dishonest" modes of construction widely prevalent in Victorian design.

Modernism after World War II
(The visual and performing arts)

In Britain and America, Modernism as a literary movement is generally considered to be relevant up to the early 1930s, and "Modernist" is rarely used to describe authors prominent after 1945. This is somewhat true for all areas of culture, with the exception of the visual and performing arts.

The post-war period left the capitals of Europe in upheaval with an urgency to economically and physically rebuild and to politically regroup. In Paris (the former center of European culture and the former capital of the art world) the climate for art was a disaster. Important collectors, dealers, and modernist artists, writers, and poets had fled Europe for New York and America. The Surrealists, and modern artists from every cultural center of Europe had fled the onslaught of the Nazis for safe haven in the United States. Many of those that didn't flee perished. A few artists, notably Pablo Picasso, Henri Matisse, and Pierre Bonnard, remained in France and survived.

The 1940s in New York City heralded the triumph of American Abstract expressionism, a modernist movement that combined lessons learned from Henri Matisse, Pablo Picasso, Surrealism, Joan Miró, Cubism, Fauvism, and early Modernism via great teachers in America like Hans Hofmann and John D. Graham. American artists benefited from the presence of Piet Mondrian, Fernand Léger, Max Ernst and the André Breton group, Pierre Matisse's gallery, and Peggy Guggenheim's gallery The Art of This Century, as well as other factors.

Pollock and Abstract Influences

During the late 1940s Jackson Pollock's radical approach to painting revolutionized the potential for all contemporary art that followed him. To some extent Pollock realized that the journey toward making a work of art was as important as the work of art itself. Like Pablo Picasso's innovative reinventions of painting and sculpture near the turn of the century via Cubism and constructed sculpture, Pollock redefined the way art gets made. His move away from easel painting and conventionality was a liberating signal to the artists of his era and to all that came after. Artists realized that Jackson Pollock's process—the placing of unstretched raw canvas on the floor where it could be attacked from all four sides using artist materials and industrial materials; drawing, staining, brushing; imagery and non-imagery—essentially blasted art making beyond any prior boundary.

Key Words

[1] modernism ['mɔdənɪzəm] n. （20世纪40年代至60年代建筑、装饰艺术等方面的）现代主义

[2] De Stijl [də'staɪl] n. [荷] 德斯太尔抽象画派，风格主义

[3] constructivism [kən'strʌktɪvɪzəm] n. 结构主义

[4] Zeitgeist ['zaɪtgaɪst] adj. 时代精神，时代思潮

[5] legacy ['legəsɪ] n. 1. 遗产，遗赠物 2. 遗留之物 3. 遗留问题；后遗症

[6] manifesto [mænɪ'festəʊ] （复数：manifestoes） n. 宣言

[7] antecedent [ˌæntɪ'siːdənt] （复数：antecedents） n. 经历

[8] revision [rɪ'vɪʒən] n. 1. 修订，修改 2. 修订本，修订版 3. 版次

[9] indulgence [ɪn'dʌldʒəns] n. 1. 纵容，迁就，宽容，包含 2. 放纵，纵容，沉溺，沉迷 3. 嗜好，爱好，享受 4. （天主教的）特赦，豁免，免罪

[10] ornamentation [ˌɔːnəmen'teɪʃən] n. 装饰，装饰品

[11] literary ['lɪtərərɪ] adj. 1. 文学（上）的 2. 精通文学的；爱好文学的；从事文学研究（或写作）的 3. 适于文学作品的；有典型文学作品特征的

[12] urgency ['ɜːdʒənsɪ] n. 1. 紧迫；急迫；急事 2. 催促；坚持

[13] onslaught ['ɔnslɔːt] n. 猛攻，攻击；冲击

[14] triumph ['traɪʌmf] n. 1. 胜利，成功 2. 巨大的成就，巨大成功；重大成就；伟大胜利 3. （巨大成功或胜利的）心满意足，喜悦，狂喜 4. （成功的）典范，楷模 vi. 1. 获胜，得胜；克服 2. 打败；战胜；成功

[15] sculpture ['skʌlptʃə] n. 1. 雕刻，雕塑，雕像 2. 雕刻品 3. 雕刻术；雕塑术 vt. & vi. 雕刻，雕塑

Key Sentences

1. The Modernist legacy is considerable in terms of design (whether furniture, tableware, textiles, lighting, advertising, and typography or other everyday things), architecture (whether public or private housing, cinemas, office blocks, and corporate headquarters), or writings (theories, manifestos, books, periodicals, and criticism). This has done much to cement Modernism firmly into the mainstream history of design.

在设计(不论是家具、餐具、纺织品、灯饰、广告、印刷品或其他日常用品)、建筑(不论是公共或私人住宅、电影院、办公大楼或公司总部),或者文字作品(理论、宣言、书籍、期刊及评论)方面的现代主义遗产是相当可观的,这一切巩固了现代主义的地位,使其成为设计史的主流。

2. Such trends found expression in the increasingly rich and vibrant vocabulary of Postmodernism, echoed in the increasingly ephemeral lifestyle enjoyed by those in the industrial world with greater levels of disposable income.

这种趋势表现在日益增长的富有的和活跃的后现代主义语言表达方式,也反映在工业社会那些获得更高收入水平的人们所享受的流行一时的生活方式。

3. In Britain and America, Modernism as a literary movement is generally considered to be relevant up to the early 1930s, and "Modernist" is rarely used to describe authors prominent after 1945.

在英国和美国,现代主义作为一种文学运动,通常被认为发生于20世纪30年代初,同时现代主义者在1945年后很少被用于描述著名作者。

课文翻译

现代主义

现代主义,也被称为现代运动,以与过去有意识的分裂为标志,它是20世纪设计实践、产品和理论出众的表达方式之一,并且以经常使用现代材料如钢管和玻璃达到视觉效果为特性,主要运用抽象的形式、空间和灯光,以及有节制地使用颜色,多以白、灰白、灰、黑为主色。在随后的1908年,阿道夫·鲁斯写了一篇著名的文章《装饰和犯罪》,再之后勒·柯布西埃用他的言论"拙劣的作品总是装饰丰富"来回应,其观点为外观通常应该是朴素的。装饰被限制使用并且与风格派和构成主义联系紧密的先锋艺术家的抽象审美原则相协调。与现代主义紧密联系的格言是"形式追随功能"。但实际上,这经常是象征意义多于实践案例,是时代精神的视觉隐喻。尽管如此,现代主义发现了物质表达的形式以及令人兴奋的,新的快速发展的交通和通信形式,新的生产方式和材料,技术和科学变化,以及由电力供应的当代生活方式。

现代主义起源于19世纪设计改良运动,并在第一次世界大战期间在德国发展。在设计(不论是家具、餐具、纺织品、灯饰、广告、印刷品或其他日常用品)、建筑(不论是公共或私人住宅、电影院、办公大楼或公司总部),或者文字作品(理论、宣言、书籍、期刊及评论)方面的现代主义遗产是相当可观的,这一切巩固了现代主义的地位,使其成为设计史的主流。此外,全世界许多博物馆大量展出从19世纪末直到20世纪最后三十年的设计藏品,这些藏品是现代主义的代表,围绕现代主义者的审美和其当下的先行者。

第二次世界大战之后,国际风格被许多主要的跨国公司所接受,用于建筑、室内、家具和室内陈设,以及办公室和陈列室的设备等。通过鲜明的现代形象提升其自身,在全球经济中树立高效、现代化和国际化的重要机构的形象。在某些人眼中,带有社会民主理想的现代主义初期的组织已经发生变化,转变为后来支持资本家。在20世纪50年代和60年代期间,设计和建筑被广泛应用,这种风格上的显著特点也引发了来自年轻一代设计师和建筑师的批评。

这种趋势表现在日益增长的富有的和活跃的后现代主义语言表达方式，也反映在工业社会那些获得更高收入水平的人们所享受的流行一时的生活方式。在 20 世纪 50 年代和 60 年代，好的设计通常指的就是现代主义美学，但是与两次战争期间的第一代现代主义者所推进的社会乌托邦基础无关。在英国，这类项目得到了由国家资金支持的工业设计理事会的许可，反对美国设计中固有的复杂造型和陈旧内涵，这对英国消费者来说具有吸引力。同时，在美国纽约的现代艺术博物馆，它们也被视为欧洲束缚下的典范。

在定义现代主义的重要角色的关键文字是尼古拉斯·佩夫斯纳的书，该书被广泛阅读，首次作为现代运动的先锋而出版。随后，它经历了重要的修订（包括在 1949 年由纽约现代艺术博物馆支持的一次）并且以现代设计先锋的标题大量重新印刷：从威廉·莫里斯到瓦特·格罗皮乌斯。佩夫斯纳提供了一种方法的报告，在这种方法里，约翰·拉斯金、威廉·莫里斯和工艺美术运动的倡导者反对历史装饰物中的道德颓废和物欲放纵，对材料不适当的应用，和广泛流行于维多利亚设计中的不诚实的结构方式。

第二次世界大战后的现代主义
（视觉和表演艺术）

在英国和美国，现代主义作为一种文学运动，通常被认为发生于 20 世纪 30 年代初，1945 年后杰出的作者很少被称为现代主义者。这在除视觉和表演艺术的各个文化领域多少是有些真实的。

战争后期的欧洲资本主义面临着巨变，经济和物质急需重建，政府急需重组。在巴黎，（过去的欧洲文化中心和艺术世界的首都），其气氛对于艺术就是一场灾难。重要的收藏家、商人和现代主义者、艺术家、作家和诗人已经逃到美国纽约。来自每个欧洲文化中心的超现实主义者和现代主义艺术家也已经逃离了纳粹的攻击，到了美国这个避风港。许多没有逃离的艺术家已经死去。极少数的艺术家，像毕加索、马蒂斯、波恩等留在法国，并幸存下来。

20 世纪 40 年代的纽约预示着美国抽象表达主义的成功，一场与课程相结合的现代主义运动兴起，来自于马蒂斯、毕加索、超现实主义、米罗、立体主义、野兽派的课程和美国著名教师汉斯·霍夫曼和约翰·D. 格雷厄姆的早期现代主义教育。美国艺术家们得益于蒙德里安、雷捷、马克思·恩斯特和安德烈·布雷顿团队、马蒂斯的展览馆、20 世纪伟大艺术古根汉姆的展览馆以及其他因素。

波洛克和抽象主义的影响

20 世纪 40 年代晚期，杰克逊·波洛克的激进绘画手法使得所有他之后的当代艺术潜能得以彻底的变革。在某些程度上，波洛克意识到制作艺术工作旅程就像艺术作品本身一样重要，就像世纪交替之际，毕加索发明的彻底变革的油画和雕塑。通过立体派和结构性的雕塑，波洛克重新定义了艺术制作的方式。他从画架和常规上移开，成为接下来艺术家新时期的一个解放的信号。艺术家们意识到波洛克的过程——在地面上放置一块未制作的油画布，它能够从四个角度操作，使用艺术家们的材料和工业材料；绘画、着色、刷光；形象化或非形象化——超越边界的艺术创作是必要的。

Free Reading 1

Ludwig Mies van der Rohe

(*b* Aachen, 27 March 1886; *d* Chicago, IL, 17 Aug 1969). German architect, furniture designer and teacher, active also in the USA, With Frank Lloyd Wright, Walter Gropius and Le Corbusier, he was a leading figure in the development of modern architecture. His reputation rests not only on his buildings and projects but also on his rationally based method of architectural education.

Ludwig Mies van der Rohe (1886—1969), Germanborn American architect, was a leading exponent of the International Style. His "skin and bones" philosophy of architecture is summed up in his famous phrase "less is more".

Early Career

Mies worked in his father's stone-carving shop and at several local design firms before he moved to Berlin joining the office of interior designer Bruno Paul. He began his architectural career as an apprentice at the studio of Peter Behrens from 1908 to 1912, where he was exposed to the current design theories and to progressive German culture, working alongside Walter Gropius and Le Corbusier. Mies served as construction manager of the Embassy of the German Empire in Saint Petersburg under Behrens. His talent was quickly recognized and he soon began independent commissions, despite his lack of a formal college-level education. A physically imposing, deliberative, and reticent man, Ludwig Mies renamed himself as part of his rapid transformation from a tradesman's son to an architect working with Berlin's cultural elite, adding the more aristocratic surname "van der Rohe". He began his independent professional career designing upper class homes in traditional Germanic domestic styles.

The American Years

Forced to flee Nazi Germany in 1937, Mies went to the United States; he became an American citizen in 1944. Mies's philosophy of architecture, which was to dominate his designs in the United States, was exemplified in his revolutionary projects of 1919 and 1920-1921 for glass skyscrapers in Berlin. They were to be "new forms from the very nature of new problems." His 1922 project for a reinforced-concrete office building epitomized all the ideals of the International Style; volume rather than mass, simplicity of surface treatment with no ornamentation, and horizontal emphasis (except in tall structures). Mies stated, "Reinforced concrete structures are skeletons by nature. No gingerbread. No fortress. Columns and girders eliminate bearing walls. This is skin and bones construction."

In 1938 Mies became director of architecture of the Illinois Institute of Technology (formerly the Armour Institute), an office he held until he resumed private practice in 1958. In his brief inaugural address he stated that "true education is concerned not only with practical goals but also with values... Education must lead us from irresponsible opinion to true responsible judgment..." He ended by quoting St. Augustine, "Beauty is the splendor of Truth."

Significance and Meaning

Mies adopted an ambitious lifelong mission to create not only a new architectural style, but also a solid intellectual foundation for a new architectural language that could be used to represent the new era of technology and production. He saw a need for architecture expressive of and in harmony with his epoch, just as Gothic architecture was for an era of spiritualism. He applied a disciplined design process using rational thought to achieve his spiritual goals. He adopted the idea that architecture communicated the meaning and significance of the culture in which it exists. The self-educated Mies painstakingly studied the great philosophers and thinkers of the past and of the day to enhance his own understanding of the character and essential qualities of the times he lived in. More than perhaps any other practising pioneer of modernism, Mies used philosophy as a basis for his work. Mies' architecture was created at a high level of abstraction, and his own descriptions of his work leave much room for interpretation. Villa Tugendhat built in 1930 in Brno, in today's Czech Republic, for Fritz Tugendhat, are shown in Figure 2.7.

Figure 2.7　Villa Tugendhat

Free Reading 2

Le Corbusier

His Life

Le Corbusier was born in 1887 in the Swiss watchmaking town of La Chaux de Fonds. His father was a highly skilled watch enameler; his mother was a pianist and music teacher. The family was Protestant; some scholars believe they were Calvinists (Sereyni 1975: 23). At the age of fifteen, Corbusier enrolled at the local trade school, L'Ecole D'Art, in order to learn the craft of watch case engraving. Corbusier's mentor at the school was Charles L'Eplattenier. L'Eplattenier's personal mission at L'Ecole was to find the most promising students alternate careers in the fine arts. He knew that eventually the craft works at La Chaux de Fonds would be replicated by machine at a cheaper price, thus destroying the local economy.

L'Eplattenier saw promise in the young Corbusier. In fact, he decreed that the young man should become an architect. Corbusier was at first ambivalent, preferring a career as a painter, but later he came to embrace the architecture profession. Under L'Eplattenier's tutelage, Corbusier was exposed to William Morris, John Ruskin, Plato and Pythagorus. Other early influences were Edward Schure's Les grands initiés and Owen Jones's Grammar of Ornament. Plato, Schure, and Jones, appear to be the most influential on Corbusier's developing worldview. From Plato, Corbusier extracted the seemingly universal ideas of Beauty, Truth, and Harmony. The forms were out there, i.e. not of this world; one only had to get beneath everyday and one's own body. Intrinsic to neo-Platonist philosophy was the notion that only a few worthy "initiés", as Schure called them, could ever know their universal forms. Artistically, neo-Platonism meant a rejection of realist representations and a concentration on getting at the true nature of an object. It also implied an antagonism toward ornament of any kind. The true forms were geometric, stylized shapes and figures.

Corbusier came to reject much of his teacher's theories on the revival of traditional arts and crafts. Instead, he developed ideas about the inevitability of capitalist rationality and the aesthetic of the machine. In fact, he began to hold the spirit of capitalism, in the form of technocratic calculations and bureaucratic order, in the highest esteem. This change appears to have been inculcated in tandem with the Bauhaus School in Vienna and his association with Auguste Perret, a Parisian Engineer. Under Perret's guidance, Corbusier learned the aesthetics of functionalism (the beauty of a carefully calculated structure sans ornament) and the positivism of the modern age. Perret was so optimistic about the new age of progress, he proclaimed, "Wars are over! There are no more frontiers!" after a successful airplane flight across the English Channel.

Corbusier and Modernity

Corbusier shared Perret's confidence and enthusiasm for the modern age. He envisaged a new and unique role for the artist/architect and the city planner that closely adhered to the capitalist spirit. Put simply, Corbusier's initial encounter with the large complex city of Paris convinced him of the need for modern housing and a modern city. Partly, this was a response to what he called the chaos around him—the enormous amount of traffic and the squalor of the industrial workers' housing. He compared this disorder to the discipline and authority of the factory and found the city lacking. Corbusier believed that the only way to impede a worker revolution was to formulate a machine for living, a dwelling that would bring the worker's home life in line with the discipline of the factory. To this end, he created the Domino housing concept, which was a rectangular structure with only four load bearing reinforced concrete members. The walls, then, could be opened up to sunlight via wrap around glass windows. The housing was purported, by Corbusier, to be a cheap, efficient way to house workers that would provide a modern ethos.

It was not just that Corbusier believed in the uplift theory of architecture, i. e., the assumption that "improved" housing would lift workers out of their culture of poverty. He also subscribed to the theory of architecture as control and discipline. Stuart Ewen, in his book *All Consuming Images*, notes that many modern thinkers presumed a correlation between the masses' behavior and architectural structures. He quotes Charlotte Bronte, famous novelist, on the Crystal Palace, "Yesterday I went for the second time to the Crystal Palace... the multitude filling the great aisles seems ruled and subdued by some invisible influence..." Corbusier was very interested in exploiting the "invisible influence" of architecture in the modern age, "The problem of the house is a problem of the epoch. The equilibrium of society today depends upon it. Architecture has for its first duty, in this period of renewal, that of bringing about a revision of values, a revision of the constituent elements of the house. We must create the mass production spirit" Later, he tersely states. "Architecture or Revolution. Revolution can be avoided" . To this end, Corbusier rationalized the house.

The historic city, then, was seen as fomenting revolution. The old "decrepit" structures from the past had to be cleared away, according to Corbusier, if the modern age was to fulfill its true duty—unlimited production of human needs and wants (progress as promised). Corbusier's first attempt at city planning came in the form of the Contemporary City Plan for Three Million People, followed by the Voisin Plan, which was application of the Contemporary City to Paris. In these early theories, he attempted to illuminate how his plan would be beneficial to business sector of the city. This was before his disillusionment with capitalism. Without going into great detail, the Contemporary City was based upon clearance of most of the Parisian landscape (a few historic monuments were to be kept), and the erection of twenty four steel and glass skyscrapers that would house the business and artistic elite. The workers were placed at the edges of the city in modern apartment structures, based on the Domino, close to their workplace—the factory. Most of the land, around eighty-five percent, was left to natural landscapes and playgrounds.

Corbusier assumed that the plan would garner support from capitalists interested in arresting the workers' movements and instituting a factory-like discipline onto the whole of society. No one took him up on it. With the depressions of the late 1920s and a tepid reception from the industrials, Corbusier lost his faith in capitalism as the ultimate bearer of progress, at least in this stage of its evolution: the plans were sound, the capitalists were too immature to realize their validity.

In 1930, Corbusier joined the syndicalist movement. Syndicalism had come to embrace, in France, an intense abhorrence of parliamentary democracy an appreciation of workers' rights—elements of the extreme left and the extreme right. Democracy was seen as a chaotic, inefficient way of regulating capitalist production. A more harmonious and more disciplined authority was created, in theory, based upon syndicats. Syndicats were groups of workers in a particular trade that elected their "natural" leader to a regional trade council. From the regional council, the most able individual was chosen to represent the regionals at the national council. The pyramid like conception reached an apex with the "natural" elites making dispassionate, scientific plans on how and what the factories should produce. For Corbusier, this meant that capitalism would have a plan and thus, would be ordered and harmonious. No longer would factories be able to overproduce and create depression; it would all be regulated from above by *Les grands initiés*.

Forays into Urbanism

For a number of years French officials had been unsuccessful in dealing with the squalor of the growing Parisian slums, and Le Corbusier sought efficient ways to house large numbers of people in response to the urban housing crisis. He believed that his new, modern architectural forms would provide a new organisational solution that would raise the quality of life of the lower classes. His Immeubles Villas (1922) was such a project that called for large blocks of cell-like individual apartments stacked one on top of the other, with plans that included a living room, bedrooms, and kitchen, as well as a garden terrace.

Not merely content with designs for a few housing blocks, soon Le Corbusier moved into studies for entire cities. In 1922, he also presented his scheme for a "Contemporary City" for three million inhabitants (Ville Contemporaine). The centerpiece of this plan was the group of sixty-storey, cruciform skyscrapers, and steel-framed office buildings encased in huge curtain walls of glass. These skyscrapers were set within large, rectangular park-like green spaces. At the very middle was a huge transportation centre that on different levels included depots for buses and trains, as well as highway intersections, and at the top, an airport. He had the fanciful notion that commercial airliners would land between the huge skyscrapers. Le Corbusier segregated pedestrian circulation paths from the roadways and glorified the use of the automobile as a means of transportation. As one moved out from the central skyscrapers, smaller low-storey, zigzag apartment blocks set far back from the street amid green space, housed the inhabitants. Le Corbusier hoped that politically-minded industrialists in France would lead the way with their efficient Taylorist and Fordist strategies adopted from American industrial models to reorganise society. As Norma Evenson has put it, "the proposed city appeared to some an audacious and compelling vision of a brave new world, and to others a frigid megalomaniac ally scaled negation of the familiar urban ambient."

In this new industrial spirit, Le Corbusier contributed to a new journal called *L' Esprit Nouveau* that advocated the use of modern industrial techniques and strategies to transform society into a more efficient environment with a higher standard of living on all socioeconomic levels. He forcefully argued that this transformation was necessary to avoid the spectre of revolution that would otherwise shake society. His dictum "Architecture or Revolution", developed in his articles in this journal, became his rallying cry for the book *Vers une architecture* (*Toward an Architecture*, previously mistranslated into English as *Towards a New Architecture*), which comprised selected articles he contributed to *L'Esprit Nouveau* between 1920 and 1923.

Theoretical urban schemes continued to occupy Le Corbusier. He exhibited his *Plan Voisin*, sponsored by another famous automobile manufacturer, in 1925. In it, he proposed to bulldoze most of central Paris, north of the Seine, and replace it with his sixty-story cruciform towers from the Contemporary City, placed in an orthogonal street grid and park-like green space. His scheme was met with only criticism and scorn from French politicians and industrialists, although they were favourable to the ideas of Taylorism and Fordism underlying Le Corbusier designs. Nonetheless, it did provoke discussion concerning how to deal with the cramped, dirty conditions that enveloped much of the city.

In the 1930s, Le Corbusier expanded and reformulated his ideas on urbanism, eventually publishing them in *La Ville radieuse* (The Radiant City) of 1935. Perhaps the most significant difference between the Contemporary City and the Radiant City is that the latter abandons the class-based stratification of the former; housing is now assigned according to family size, not economic position. *La Ville radieuse* also marks Le Corbusier's increasing dissatisfaction with capitalism and his turn to the right-wing syndicalism of Hubert Lagardelle. During the Vichy regime, Le Corbusier received a position on a planning committee and made designs for Algiers and other cities. The central government ultimately rejected his plans, and after 1942 Le Corbusier withdrew from political activity.

After World War II, Le Corbusier attempted to realize his urban planning schemes on a small scale by constructing a series of "unités" (the housing block unit of the Radiant City) around France. The most famous of these was the Unité d' Habitation of Marseilles (1946—1952). In the 1950s, a unique opportunity to translate the Radiant City on a grand scale presented itself in the construction of Chandigarh, the new capital for the Indian states of Punjab and Haryana. Le Corbusier was brought on to develop the plan of Albert Mayer.

Corbusier's Radiant City

The Radiant City grew out of this new conception of capitalist authority and a pseudo-appreciation for workers' individual freedoms. The plan had much in common with the Contemporary City—clearance of the historic cityscape and rebuilding utilizing modern methods of production. In the Radiant City, however, the pre-fabricated apartment houses, *les unites*, were at the center of "urban" life. *Les unites* were available to everyone (not just the elite) based upon the size and needs of each particular family. Sunlight and recirculating air were provided as part of the design. The scale of the apartment houses was fifty meters high, which would accommodate, according to Corbusier, 2,700 inhabitants with fourteen square meters of space per person. The building would be placed upon pilotus, five meters off the ground, so that more land could be given over to nature. Setback from other unites would be achieved by *les redents*, patterns that Corbusier created to lessen the effect of uniformity.

Inside *les unites* were the vertical streets, i. e. the elevators, and the pedestrian interior streets that connected one building to another. As in the Contemporary City, corridor streets were destroyed. Automobile traffic was to circulate on pilotus supported roadways five meters above the earth. The entire ground was given as a "gift" to pedestrians, with pathways running in orthogonal and diagonal projections. Other transportation modes, like subways and trucks, had their own roadways separate from automobiles. The business center, which had engendered much elaboration in the Contemporary City, was positioned to the north of *les unites* and consisted of Cartesian (glass & steel) skyscrapers every 400 meters. The skyscrapers were to provide office space for 3,200 workers per building.

Corbusier spends a great deal of the Radiant City manifesto elaborating on services available to the residents. Each apartment block was equipped with a catering section in the basement, which would prepare daily meals (if wanted) for every family and would complete each families' laundry chores. The time saved would enable the individual to think, write, or utilize the play and sports grounds which covered much of the city's land. Directly on top of the apartment houses were the roof top gardens and beaches, where residents sun themselves in Anatural surroundings—fifty meters in the air. Children were to be dropped off at *les unites'* day care center and raised by scientifically trained professionals. The workday, so as to avoid the crisis of overproduction, was lowered to five hours a day. Women were enjoined to stay at home and perform household chores, if necessary, for five hours daily. Transportation systems were also formulated to save the individual time. Corbusier bitterly reproaches advocates of the horizontal garden city (suburbs) for the time wasted commuting to the city. Because of its compact and separated nature, transportation in the Radiant City was to move quickly and efficiently. Corbusier called it the vertical garden city.

Many scholars have adopted the notion that the Corbusier of the Radiant City was a kinder, gentler Corbusier. However, they have failed to consider that the so-called individual freedoms that Corbusier promoted were not freedoms at all. Certainly, Corbusier provided leisure time activities that he enjoyed, such as sunbathing on the roof or playing basketball. But, are these pastimes necessarily freedom? Corbusier's individuals were not allowed to have a voice in the governance of their lives; they are able to behave, but not to act. Additionally, there is no room in the Radiant City for individuals to act nonrationally. The leisure time advocated by Corbusier is one filled with healthy "day minded" pursuits. There can be no extravagance or chaotic excess. The town lunatic would have to go the way of ninety-nine percent of the historic city. Indeed, it is improbable that ninety-nine percent of humanity will ever behave in so-called rational ways. Thus, Corbusier's vision suffers from an naive conception of human nature.

But, this is not the main problem with his thesis for the Radiant City. Quite simply, his notion of authority is both patriarchal and bureaucratic, what Richard Sennett refers to as the authority of false love and the authority of no love (Sennett 1980). Corbusier maintained, following Plato and Schure, that universal truth, beauty, and goodness could be ascertained by those who had divorced themselves from matter (human bodies). *Les grands initiés* could then prescribe a plan grounded in objective calculations and scientific facts. There could be no debate, i. e. no politics regarding the precepts of the plan. Humanity was to accept this discipline as a necessary, objective ordering of reality by a doting, paternalistic authority. Corbusier put it like this, "Authority must step in, patriarchal authority, the authority of a father concerned for his children" (Le Corbusier 1967: 152).

The Modulor

Le Corbusier explicitly used the golden ratio in his Modulor system for the scale of architectural proportion. He saw this system as a continuation of the long tradition of Vitruvius, Leonardo da Vinci's "Vitruvian Man", the work of Leon Battista Alberti, and others who used the proportions of the human body to improve the appearance and function of architecture. In addition to the golden ratio, Le Corbusier based the system on human measurements, Fibonacci numbers, and the double unit. He took Leonardo's suggestion of the golden ratio in human proportions to an extreme: he sectioned his model human body's height at the navel with the two sections in golden ratio, then subdivided those sections in golden ratio at the knees and throat; he used these golden ratio proportions in the Modulor system.

Le Corbusier's 1927 Villa Stein in Garches exemplified the Modulor system's application. The villa's rectangular ground plan, elevation, and inner structure closely approximate golden rectangles.

Le Corbusier placed systems of harmony and proportion at the centre of his design philosophy, and his faith in the mathematical order of the universe was closely bound to the golden section and the Fibonacci series, which he described as "rhythms apparent to the eye and clear in their relations with one another. And these rhythms are at the very root of human activities. They resound in Man by an organic inevitability, the same fine inevitability which causes the tracing out of the Golden Section by children, old men, savages, and the learned."

Furniture

Corbusier said, "Chairs are architecture, sofas are bourgeois." Le Corbusier began experimenting with furniture design in 1928 after inviting the architect, Charlotte Perriand, to join his studio. His cousin, Pierre Jeanneret, also collaborated on many of the designs. Before the arrival of Perriand, Le Corbusier relied on ready-made furniture to furnish his projects, such as the simple pieces manufactured by Thonet, the company that manufactured his designs in the 1930s.

In 1928, Le Corbusier and Perriand began to put the expectations for furniture Le Corbusier outlined in his 1925 book *L'Art Décoratif d'aujourd'hui* into practice. In the book he defined three different furniture types: *type-needs*, *type-furniture*, and *human-limb objects*. He defined human-limb objects as "Extensions of our limbs and adapted to human functions that are type-needs and type-functions, therefore type-objects and type-furniture. The human-limb object is a docile servant. A good servant is discreet and self-effacing in order to leave his master free. Certainly, works of art are tools, beautiful tools. And long live the good taste manifested by choice, subtlety, proportion, and harmony".

The first results of the collaboration were three chrome-plated tubular steel chairs designed for two of his projects, The Maison la Roche in Paris and a pavilion for Barbara and Henry Church. The line of furniture was expanded for Le Corbusier's 1929 Salon d'Automne installation, *Equipment for the Home*.

In the year 1964, while Le Corbusier was still alive, Cassina S. p. A. of Milan acquired the exclusive worldwide rights to manufacture his furniture designs. Today many copies exist, but Cassina is still the only manufacturer authorised by the *Fondation Le Corbusier*.

Five Points of Architecture

It was Le Corbusier's Villa Savoye (1929—1931) that most succinctly summed up his five points of architecture that he had elucidated in the journal *L'Esprit Nouveau* and his book *Vers une architecture*, which he had been developing throughout the 1920s. First, Le Corbusier lifted the bulk of the structure off the ground, supporting it by *pilotis*—reinforced concrete stilts. These *pilotis*, in providing the structural support for the house, allowed him to elucidate his next two points: a free façade, meaning non-supporting walls that could be designed as the architect wished, and an open floor plan, meaning that the floor space was free to be configured into rooms without concern for supporting walls. The second floor of the Villa Savoye includes long strips of ribbon windows that allow unencumbered views of the large surrounding yard, and which constitute the fourth point of his system. The fifth point was the Roof garden to compensate the green area consumed by the building and replacing it on the roof. A ramp rising from the ground level to the third floor roof terrace, allows for an architectural promenade through the structure. The white tubular railing recalls the industrial "ocean-liner" aesthetic that Le Corbusier much admired. As if to put an exclamation point on Le Corbusier's homage to modern industry, the driveway around the ground floor, with its semicircular path, measures the exact turning radius of a 1927 Citroën automobile.

Major Buildings and Projects (See Figure 2.8, Figure 2.9, Figure 2.10)

Figure 2.8　National Museum of Western Art in Tokyo, Japan

Figure 2.9　Centre Le Corbusier (Heidi Weber Museum) in Zürich-Seefeld (Zürichhorn)

Figure 2.10　The Open Hand Monument is one of numerous projects in Chandigarh, India designed by Le Corbusier

Free Reading 3

Form Follows Function

Form follows function is a principle associated with modern architecture and industrial design in the 20th century. The principle is that the shape of a building or object should be primarily based upon its intended function or purpose.

In the context of design professions *form follows function* seems like good sense but on closer examination it becomes problematic. Linking the relationship between the form of an object and its intended purpose is a good idea for designers and architects, but it is not always by itself a complete design solution. Defining the precise meaning(s) of the phrase "form follows function" opens a discussion of design integrity that remains an important, lively debate.

Origins of the Phrase

The authorship of the phrase is often ascribed to the American sculptor Horatio Greenough, whose thinking to a large extent predates the later functionalist approach to architecture. It was, however, the American architect Louis Sullivan who coined the phrase, in 1896, in his article *The Tall Office Building Artistically Considered*. Here Sullivan actually said "form ever follows function", but the simpler phrase is the one usually remembered. For Sullivan this was distilled wisdom, an aesthetic credo, the single "rule that shall permit of no exception". The full quote is thus:

 It is the pervading law of all things organic and inorganic,
  Of all things physical and metaphysical,
  Of all things human and all things super-human,
  Of all true manifestations of the head,
  Of the heart, of the soul,
  That the life is recognizable in its expression,
  That form ever follows function. This is the law.

Sullivan developed the shape of the tall steel skyscraper in late 19th century Chicago at the very moment when technology, taste and economic forces converged violently and made it necessary to drop the established styles of the past. If the shape of the building wasn't going to be chosen out of the old pattern book something had to determine form, and according to Sullivan it was going to be the purpose of the building. It was "form follows function", as opposed to "form follows precedent". Sullivan's assistant Frank Lloyd Wright adopted and professed the same principle in slightly different form—perhaps because shaking off the old styles gave them more freedom and latitude.

Is Ornamentation "Functional"?

In 1908 the Austrian architect Adolf Loos famously proclaimed that architectural ornament was criminal, and his essay on that topic would become foundational to Modernism and eventually trigger the careers of Le Corbusier, Walter Gropius, Alvar Aalto, Mies van der Rohe and Gerrit Rietveld. The Modernists adopted both of these equations—form follows function, ornament is a crime—as moral principles, and they celebrated industrial artifacts like steel water towers as brilliant and beautiful examples of plain, simple design integrity.

These two principles—form follows function, ornament is crime—are often invoked on the same occasions for the same reasons, but they do not mean the same thing. If ornament on a building may have social usefulness like aiding wayfinding, announcing the identity of the building, signaling scale, or attracting new customers inside, then ornament can be seen as functional.

Conversely the argument "ornament is crime" doesn't say anything about function. It is an aesthetic preference inspired by the Machine Age. Ornament becomes an unnecessary relic, or worse, an impediment to optimal engineering design and equipment maintenance. Other stylistic "non-functional" features may rest untouched (e.g., the feeling of space, the composition of the volumes) as we can see in the subsequent abstracted and non-ornamented styles. Much of the confusion between these two concepts comes from the fact that ornament traditionally derives from a function becoming a stylistic character.

Core Text 6

Bauhaus

Bauhaus, a "comprehensive" art school, which was founded in Weimar in 1919 with official support (Staatliches Bauhaus) under the direction of W. Gropius (1883—1969). Bauhaus ("House of Building" or "Building School") was a school in Germany that combined crafts and the fine arts, and was famous for the approach to design that it publicized and taught. It operated from 1919 to 1933.

The Bauhaus revolutionized art training by combining the teaching of the pure arts with the study of crafts. Philosophically, the school was built on the idea that design did not merely reflect society; it could actually help to improve it. The teaching plan insisted on functional craftsmanship in every field, with a concentration on the industrial problems of mechanical mass production. Bauhaus style was characterized by economy of method, a severe geometry of form, and design that took into account the nature of the materials employed. The school's concepts aroused vigorous opposition from right-wing politicians and academicians.

In 1925 the Bauhaus moved to the more friendly atmosphere of Dessau, where Gropius designed special buildings to house the various departments. Gropius resigned in 1928, and the leadership was continued by the architect Hannes Meyer, who in turn was replaced in 1930 by Ludwig Mies van der Rohe. In the summer of 1932 opposition to the school had increased to such an extent that the city of Dessau withdrew its support. (The Bauhaus Dessau shown in Figure 2.11). The school was then moved to Berlin, where the faculty endeavored to carry on their ideas, but in 1933 the Nazi government closed the school entirely. The Bauhaus ideas, enveloping design in architecture, furniture, weaving, and typography, among others, had by this time found wide acclaim in many parts of the world and especially in the United States. Gropius himself went to the United States and taught at Harvard, exercising considerable influence. The Chicago Institute of Design, founded by Moholy-Nagy, most completely carried on the teaching plan of the Bauhaus.

Bauhaus style became one of the most influential currents in Modernist architecture and modern design. The Bauhaus had a profound influence upon subsequent developments in art, architecture, graphic design, interior design, industrial design, and typography.

Figure 2.11 The Bauhaus Dessau

Bauhaus and German Modernism

Defeat in World War I, the fall of the German monarchy and the abolition of censorship under the new, liberal Weimar Republic allowed an upsurge of radical experimentation in all the arts, previously suppressed by the old regime. Many Germans of left-wing views were influenced by the cultural experimentation that followed the Russian Revolution, such as constructivism. Such influences can be overstated: Gropius himself did not share these radical views, and said that Bauhaus was entirely apolitical. Just as important was the influence of the 19th century English designer William Morris, who had argued that art should meet the needs of society and that there should be no distinction between form and function. Thus the Bauhaus style, also known as the international style, was marked by the absence of ornamentation and by harmony between the function of an object or a building and its design.

However, the most important influence on Bauhaus was modernism, a cultural movement whose origins lay as far back as the 1880s, and which had already made its presence felt in Germany before the World War, despite the prevailing conservatism. The design innovations commonly associated with Gropius and the Bauhaus—the radically simplified forms, the rationality and functionality, and the idea that mass-production was reconcilable with the individual artistic spirit—were already partly developed in Germany before the Bauhaus was founded. The German national designers' organization Deutscher Werkbund was formed in 1907 by Hermann Muthesius to harness the new potentials of mass production, with a mind towards preserving Germany's economic competitiveness with England. In its first seven years, the Werkbund came to be regarded as the authoritative body on questions of design in Germany, and was copied in other countries. Many fundamental questions of craftsmanship vs. mass production, the relationship of usefulness and beauty, the practical purpose of formal beauty in a commonplace object, and whether or not a single proper form could exist, were argued out among its 1,870 members (by 1914).

The entire movement of German architectural modernism was known as Neues Bauen. Beginning in June 1907, Peter Behrens' pioneering industrial design work for the German electrical company AEG successfully integrated art and mass production on a large scale. He designed consumer products, standardized parts, created clean-lined designs for the company's graphics, developed a consistent corporate identity, built the modernist landmark AEG Turbine Factory, and made full use of newly developed materials such as poured concrete and exposed steel. Behrens was a founding member of the Werkbund, and both Walter Gropius and Adolf Meier worked for him in this period.

When the German zeitgeist ("spirit of the times") had turned from emotional expressionism to the matter-of-fact new objectivity, an entire group of working architects, including Erich Mendelsohn, Bruno Taut and Hans Poelzig, turned away from fanciful experimentation, and turned toward rational, functional, sometimes standardized building. Beyond the Bauhaus, many other significant German-speaking architects in the 1920s responded to the same aesthetic issues and material possibilities as the school. They also responded to the promise of a "minimal dwelling" written into the new Weimar Constitution. Ernst May, Bruno Taut, and Martin Wagner, among others, built large housing blocks in Frankfurt and Berlin. The acceptance of modernist design into everyday life was the subject of publicity campaigns, well-attended public exhibitions like the Weissenhof Estate, films, and sometimes fierce public debate.

Architectural Output

The paradox of the early Bauhaus was that, although its manifesto proclaimed that the ultimate aim of all creative activity was building, the school did not offer classes in architecture until 1927. The single most profitable tangible product of the Bauhaus was its wallpaper.

During the years under Gropius (1919—1927), he and his partner Adolf Meyer observed no real distinction between the output of his architectural office and the school. So the built output of Bauhaus architecture in these years is the output of Gropius: the Sommerfeld house in Berlin, the Otte house in Berlin, the Auerbach house in Jena, and the competition design for the Chicago Tribune Tower, which brought the school much attention. The definitive 1926 Bauhaus building in Dessau is also attributed to Gropius. Apart from contributions to the 1923 Haus am Horn, student architectural work amounted to unbuilt projects, interior finishes, and craft work like cabinets, chairs and pottery.

In the next two years under Meyer, the architectural focus shifted away from aesthetics and towards functionality. There were major commissions: one from the city of Dessau for five tightly designed "Laubengangh user" (apartment buildings with balcony access), which are still in use today, and another for the headquarters of the Federal School of the German Trade Unions (ADGB) in Bernau bei Berlin. Meyer's approach was to research users' needs and scientifically develop the design solution.

Mies van der Rohe repudiated Meyer's politics, his supporters, and his architectural approach. As opposed to Gropius's "study of essentials", and Meyer's research into user requirements, Mies advocated a "spatial implementation of intellectual decisions", which effectively meant an adoption of his own aesthetics. Neither van der Rohe nor his Bauhaus students saw any projects built during the 1930s.

The popular conception of the Bauhaus as the source of extensive Weimar-era working housing is not accurate. Two projects, the apartment building project in Dessau and the Törten row housing also in Dessau, fall in that category, but developing worker housing was not the first priority of Gropius or Mies. It was the Bauhaus contemporaries Bruno Taut, Hans Poelzig and particularly Ernst May, as the city architects of Berlin, Dresden and Frankfurt respectively, who are rightfully credited with the thousands of socially progressive housing units built in Weimar Germany. In Taut's case, the housing he built in south-west Berlin during the 1920s, is still occupied, and can be reached by going easily from the U-Bahn stop Onkel Toms Hütte.

Impact

The Bauhaus had a major impact on art and architecture trends in Western Europe, the United States, Canada and Israel (particularly in White City, Tel Aviv) in the decades following its demise, as many of the artists involved fled, or were exiled, by the Nazi regime. Tel Aviv, in fact, has been named to the list of world heritage sites by the UN due to its abundance of Bauhaus architecture in 2004. It had some 4,000 Bauhaus buildings erected from 1933 on. Walter, Marcel Breuer, and László Moholy-Nagy re-assembled in Britain during the mid 1930s to live and work in the Isokon project before the war caught up with them. Both Gropius and Breuer went to teach at the Harvard Graduate School of Design and worked together before their professional split. The Harvard School was enormously influential in America in the late 1920s and early 1930s, producing such students as Philip Johnson, I. M. Pei, Lawrence Halprin and Paul Rudolph, among many others.

In the late 1930s, Mies van der Rohe re-settled in Chicago, enjoyed the sponsorship of the influential Philip Johnson, and became one of the pre-eminent architects in the world. Moholy-Nagy also went to Chicago and founded the New Bauhaus school under the sponsorship of industrialist and philanthropist Walter Paepcke. This school became the Institute of Design, part of the Illinois Institute of Technology. Printmaker and painter Werner Drewes was also largely responsible for bringing the Bauhaus aesthetic to America and taught at both Columbia University and Washington University in St. Louis. Herbert Bayer, sponsored by Paepcke, moved to Aspen, Colorado in support of Paepcke's Aspen projects at the Aspen Institute. In 1953, Max Bill, together with Inge Aicher-Scholl and Otl Aicher, founded the Ulm School of Design (German: Hochschule für Gestaltung - HfG Ulm) in Ulm, Germany, a design school in the tradition of the Bauhaus. The school is notable for its inclusion of semiotics as a field of study. The school closed in 1968, but the "Ulm Model" concept continues to influence international design education.

One of the main objectives of the Bauhaus was to unify art, craft, and technology. The machine was considered a positive element, and therefore industrial and product design were important components. *Vorkurs* ("initial" or "preliminary course") was taught; this is the modern day "Basic Design" course that has become one of the key foundational courses offered in architectural and design schools across the globe. There was no teaching of history in the school because everything was supposed to be designed and created according to first principles rather than by following precedent.

One of the most important contributions of the Bauhaus is in the field of modern furniture design. The ubiquitous Cantilever chair by Dutch designer Mart Stam, using the tensile properties of steel, and the Wassily Chair designed by Marcel Breuer are two examples.

The physical plant at Dessau survived World War II and was operated as a design school with some architectural facilities by the German Democratic Republic. This included live stage productions in the Bauhaus theater under the name of *Bauhausbühne* ("Bauhaus Stage"). After German reunification, a reorganized school continued in the same building, with no essential continuity with the Bauhaus under Gropius in the early 1920s. In 1979 Bauhaus-Dessau College started to organize postgraduate programs with participants from all over the world. This effort has been supported by the Bauhaus-Dessau Foundation which was founded in 1974 as a public institution.

American art schools have also rediscovered the Bauhaus school. The Master Craftsman Program at Florida State University bases its artistic philosophy on Bauhaus theory and practice.

Key Words

[1] Bauhaus [ˈbauhaus] *n*. 包豪斯建筑学派

[2] craftsmanship [ˈkrɑːftsmənʃɪp] *n*. 技巧；技术

[3] faculty [ˈfækəltɪ] *n*. 1. 能力，才能 2. 官能；天赋 3. （高等院校的）系；院 4. （高等院校中院、系的）全体教师 5. 官能；天赋 6. （高等院校的）系，院

[4] typography [taɪˈpɒgrəfɪ] *n*. 凸版印刷术，排印；印刷样式

[5] graphic design 平面设计

[6] abolition [æbəˈlɪʃən] *n*. 1. 废除，废止 2. （常用 A-）[美史] 黑奴制度的废除，废奴运动 3. （常用 A-）[英史] 奴隶贩卖的制止，奴隶制度的废除

[7] censorship [ˈsensəʃɪp] *n*. 审查员的职权；审查（制度）

[8] usefulness [ˈjuːsfʊlnəs] *n*. 有用，有益，有效

[9] wallpaper [ˈwɔːlpeɪpə] *n*. 壁纸；墙纸 *vt*. 往（屋里墙上）贴壁纸 *vi*. 贴壁纸

[10] functionality [fʌŋkʃəˈnælətɪ] *n*. 1. 功能性，泛函性 2. 实用；符合实际 3. 设计目的；设计功能 4. （计算机或电子系统的）功能

[11] semiotics [ˌsiːmɪˈɒtɪks] *n*. 1. 符号学 2. 症状学

[12] furniture design 家具设计

[13] ubiquitous [juːˈbɪkwɪtəs] *adj*. 1. [正] 普遍存在的；似乎无处不在的；十分普遍的

[14] continuity [ˌkɒntɪˈnjuːɪtɪ] *n*. 1. 连续（性），持续（性）2. （逻辑上的）连接，联结 3. （电影或电视节目场景中服装、物体的）一致性，衔接 4. [数学] 连续性

Key Sentences

1. The Bauhaus revolutionized art training by combining the teaching of the pure arts with the study of crafts.

包豪斯通过将纯艺术理论教学和工艺研究结合，彻底革新了纯艺术教育。

2. Bauhaus style became one of the most influential currents in Modernist architecture and modern design. The Bauhaus had a profound influence upon subsequent developments in art, architecture, graphic design, interior design, industrial design, and typography.

包豪斯风格成为现代主义建筑和现代设计最具影响力的潮流。包豪斯对其后的艺术、建筑、平面设计、室内设计、工业设计和版面设计产生了深远影响。

3. One of the main objectives of the Bauhaus was to unify art, craft, and technology. The machine was considered a positive element, and therefore industrial and product design were important components. *Vorkurs* ("initial" or "preliminary course") was taught; this is the modern day "Basic Design" course that has become one of the key foun-

dational courses offered in architectural and design schools across the globe.

包豪斯的一项主要目标是将艺术、工艺和技术统一起来。机械被认为是一个积极因素，所以工业和产品设计是重要的组成部分。"Vorkurs"（"初始"或"基础教程"），如今的"基本设计"课程，已经成为全球建筑与设计学校设置的关键性基础课程之一。

课文翻译

包豪斯

包豪斯是一所综合性艺术学校，于1919年由瓦尔特·格罗佩斯（1883—1969）在官方支持下在魏玛创立。包豪斯（"建筑之家"或者"建筑学院"）是德国国立包豪斯学校的通称，它是一所将工艺和美术结合的学校，得名于其宣传和讲授设计的方式。学校从1919年运作到1933年。

包豪斯通过将纯艺术理论教学和工艺研究相结合，革新了艺术培训。从哲学上讲，学校以设计不仅仅反映社会，实际上可以促进社会发展为宗旨。教育计划始终坚持将功能性工艺应用在各个领域，集中在机械化大批量生产的工业问题上。包豪斯的风格特点是经济方法、严谨的几何形式以及设计时考虑所用材料的性质。学校的理念引起了右翼政客和学者的强烈反对。

1925年，包豪斯迁到氛围更友好的德绍，在这里，格罗佩斯为各个部门设计了特别建筑。1928年，格罗佩斯辞职，学校校长由建筑师汉斯·迈耶担当。1930年，米斯·凡德罗代替汉斯成为校长。1932年夏天，对学校的反对之声增加致使德绍撤回了对它的支持。学校后来转移到柏林，在那里，全体教师努力坚持自己的思想，但是，1933年，纳粹政府彻底地关闭了学校。包豪斯思想反映在建筑外观设计、家具、纺织和版面设计上，这一时期受到了世界许多地区，尤其是美国的广泛赞誉。格罗佩斯亲自执教于美国哈佛大学，产生了相当大的影响。芝加哥设计研究所由莫霍伊—纳吉成立，最完整地实施了包豪斯的教学计划。

包豪斯风格成为现代主义建筑和现代设计最具影响力的潮流。它对其后的艺术、建筑、平面设计、室内设计、工业设计和版面设计产生了深远影响。

包豪斯和德国现代主义

第一次世界大战失败后，德国君主立宪制垮台，在新的自由政权魏玛共和国统治下，检查制度被废除，新政权允许高涨起来的、被先前旧制度抑制的所有艺术形式积极尝试。许多德国左翼观点受到紧随俄国革命的文化尝试，如构成主义的影响。这些影响可能有些夸张，格罗佩斯本人并不同意这些激进的看法，并指出包豪斯是完全非政治性的。正如19世纪英国设计师威廉·莫里斯产生的重要影响，莫里斯称，艺术应该满足社会需要，形式和功能不应该有差别。所以，包豪斯风格，也可以称为国际主义风格，标志着没有装饰和物体或建筑物的功能与其设计之间的和谐。

然而，对包豪斯产生最重要影响的是现代主义运动，一场根源追溯到19世纪80年代的文化运动，尽管当时保守主义占主流，但在世界大战前的德国已经初见端倪。设计革新通常与格罗佩斯和包豪斯联系起来——简易的形式、合理性和功能性，还有大规模生产与个人艺术精神调和的思想——在包豪斯成立前就已经在德国部分地发展起来。德国国家设计组织——德意志制造联盟于1907年由赫尔曼·穆特修斯创立，利用大规模生产的潜力，考虑维护德国与英格兰的经济竞争。在最初的七年里，制造联盟渐渐被视为德国的设计权威机构，被其他国家效仿。许多艺术和大规模生产的基本问题，如实用和美感的关系，一件普通物品形式美的实用目的和是否能形成一种简单正确的形式问题通过了1870名成员的辩论(1914年)并得出了结论。

整个德国建筑界现代主义运动被称为"Neues Bauen"。1907年6月，彼得·贝伦斯为德国AEG电气公司设计的开创性地工业设计作品大规模地成功融合了艺术和大批量生产。他设计了消费产品和标准化部件，为公司平面设计创造了简洁的设计，并制造了持续性的企业形象，建立了现代主义的里程碑——AEG公司汽轮机厂，并充分利用了新开发的材料如混凝土和外露钢筋。贝伦斯是制造联盟的创始人之一，这一时期，瓦尔特·格罗佩斯和阿道夫·迈尔均为他工作。

包豪斯成立之时，正值德国的时代精神从情感上的表现主义到实际上的新客观物质性的转变。整个集团的建筑师，包括埃里希·门德尔松、布鲁诺·陶特和汉斯·伯尔齐格，从富于幻想的尝试试验转向理性和功能性，有时是标准化的建筑。除了包豪斯，许多其他重要的20世纪20年代的德国建筑师和学校一样反映了同样的美学问题和材料的可能性问题。他们还回应了新魏玛宪法中写入的"最小住宅"的承诺。在众多人中，恩斯特·梅尔、布鲁诺·陶特和马汀·瓦格纳在富兰克林和柏林建造了大型建筑社区。接受现代主义设计融入日常生活是公众争论的一个主题，良好互动的公共展览如国际住宅展建筑群，电影，有时会展开激烈的公开辩论。

建筑输出

早期包豪斯的矛盾在于，虽然它的宣言宣称所有创造性活动的最终目的是建造，学校1927年之前并没有开设建筑课程。包豪斯唯一最能获取利润的有形产品是它的墙纸。

在格罗佩斯领导的几年间(1919—1927年)，他和他的搭档阿道夫·迈耶注意到，办公室和学校建筑没有真正的区别。因此，包豪斯这期间建筑方面的作品实际是格罗佩斯的作品：柏林的夏日屋、柏林的"Otte house"、耶拿的"Auerbach house"、为竞赛设计的芝加哥论坛报大厦，为学校赢得了许多关注。1926年，建立在德绍的包豪斯经典建筑也归功于格罗佩斯。除了1923年的贡献号角屋，学生在建筑方面的作品累计有非建造项目，室内装修，还有诸如柜、椅子和陶器这样的工艺品。

迈耶继任两年后，建筑风格的重点由美学主义转向了功能主义，主要代表作有德绍市五个简约风格之一的"Laubenganghäuser"（露台大门住宅），至今仍在使用。另一个是贝尔瑙柏林的联邦院德国工会联合会的总部大楼。迈耶的思想是研究用户的需求和设计方法的科学发展。

米斯·凡德罗反对迈耶的政策以及他的支持者和建筑方法。同格罗佩斯的"研究要点"及迈耶的研究用户需求相反，米斯主张"智力决策的空间执行"，这实际上意味着对他自己美学的采纳。凡德罗和他的包豪斯学生在20世纪30年代都没有任何项目。

包豪斯作为魏玛时期工作室的广泛来源这个流行的说法是不准确的。两个大项目——德绍的公寓楼项目和同样在德绍的托腾社区属于上述此类，但是开发工人住房的既不是先前的格罗佩斯，也不是迈耶，而是包豪斯的当代人物布鲁诺·陶斯特，汉斯·伯尔齐格，恩斯特·梅尔。他们各自作为柏林、德累斯顿和法兰克福的城市建筑师，在德国魏玛创造了数以千计具有社会进步意义的住宅。在陶斯特案例中，他在20世纪20年代于柏林西南部建造的房屋仍在使用中，从汤姆斯地铁站很容易到达那里。

影响

在包豪斯结束的几十年里，它对西欧、美国、加拿大和以色列（尤其是特拉维夫白城）的艺术和建筑发展倾向产生了主要影响，许多艺术家在纳粹政权的压迫下逃亡或者被流放。2004年，特拉维夫因其丰富的包豪斯建筑被联合国列入世界遗产名单。1933年以来，那里大约有4000座包豪斯建筑物。20世纪30年代在战争侵袭之前，沃尔特·格罗佩斯、马塞尔·布鲁尔、拉斯洛莫霍伊—纳吉重新聚集在英国在"Isokon"项目工作。格罗佩斯和布鲁尔都去哈佛大学设计研究所教书，并在专业解散前共同工作。哈佛大学在20世纪20年代后期和30年代初期在美国产生了巨大的影响力，培养了如菲利普·约翰逊、贝聿铭、劳伦斯·哈普林和保罗·鲁道夫这样的学生。

20世纪30年代末期，米斯·凡德罗再次在芝加哥定居，受到颇具影响力的菲利普·约翰逊的资助，被称为世界上卓越的建筑师。莫霍伊—纳吉也前往芝加哥，并在实业家、慈善家沃尔特·佩普吉的赞助下建立了新包豪斯学校。这所学校成为伊利诺理工大学设计学院。版画印刷者、画家维尔纳·德雷维斯也对包豪斯美学进入美国做了大量贡献，并且同时在哥伦比亚大学和圣路易斯华盛顿大学教学。赫伯特·拜耳受到佩普吉的赞助，转移到科罗拉多的阿斯本，来支持阿斯本研究所佩普吉的项目。1953年马克思·比尔和英格·艾舍·绍尔，奥托·艾舍一起在德国乌尔姆创立了乌尔姆设计学院，这是一所沿袭了包豪斯传统思想的设计学校。学校的著名之处是将符号学纳入研究领域。学校于1968年关闭，但是"乌尔姆模式"的概念继续影响着国际设计教育。

包豪斯的一项主要目标是将艺术、工艺和技术统一起来。机械被认为是一个积极因素，所以工业和产品设计是重要的组成部分。"Vorkurs"（"初始"或"基础教程"），如今的"基本设计"课程，已经成为全球各地建筑和设计学校均设置的一项关键性的基础课程之一。学校没有任何教学历史，设计和创造都是根据最初原则，而不是效仿先例。

包豪斯最重要的一项贡献是现代家具领域。由荷兰设计师沃尔玛·斯塔姆设计的无处不在的悬臂椅，利用了钢材的拉伸性能，以及马塞尔·布鲁尔设计的瓦西里椅子是两个重要例子。

第二次世界大战期间，德绍的工厂设施幸存下来，一些建筑场所被德意志民主共和国作为设计学校来运作，包括"*Bauhausbühne*"名义下的包豪斯剧院的生活舞台作品。德国统一后，在这个建筑上重新建立了一所学校，与20世纪20年代格罗佩斯领导下的包豪斯没有必然的联系。1979年，包豪斯德绍学校开始组织面向全世界的研究生课程。这项努力得到了于1974年成立的，作为一个公共机构出现的包豪斯德绍基金会的支持。

美国艺术学校也重新开发了包豪斯学校。在佛罗里达州立大学的硕士工艺师课程研究以包豪斯艺术哲学的理论与实践为基础。

Free Reading 1

Walter Gropius

The German-American architect, educator, and designer Walter Gropius（1883—

1969) was director of the famed Bauhaus in Germany from 1919 to 1928 and occupied the chair of architecture at the Harvard University Graduate School of Design from 1938 to 1952.

Walter Gropius was born in Berlin on May 18, 1883. Although he studied architecture in Berlin and Munich (1903—1907), he received no degree. He then went to work in Berlin for Peter Behrens, one of several German architects who was influenced by the British Arts and Crafts movement and who attempted to go further by adapting good design to machine production.

In 1910 Gropius set up practice with Adolf Meyer. They designed the Fagus Works in Alfeld an der Leine (1911) and the office building at the Werkbund Exhibition in Cologne (1914), using a combination of masonry and steel construction, from which, in some areas, the external glass sheathing was hung. The plan of the Cologne building was axially designed in the Beaux-Arts tradition, but the major influence was predominantly that of Frank Lloyd Wright, whose "prairie houses" were widely known in Europe through the 1910 and 1911 publications of Ernst Wasmuth in Berlin. Gropius and Meyer were influenced by Wright's style especially in the horizontality and the wide overhanging eaves, but also in the symmetry, the corner pavilions, and the whole spirit of Wright's concept. World War I interrupted their architectural practice, and thereafter they designed only one project prior to Meyer's death in 1924: the unsuccessful entry for the Chicago Tribune Tower competition of 1922.

The Bauhaus

During the war Gropius was invited to become the director of the Grand Ducal Saxon School of Applied Arts and the Saxon Academy of Fine Arts in Weimar, and he took up his duties at war's end. He combined the two schools into the Staatliches Bauhaus (State Building House) in 1919. The aim of the Bauhaus was a "unity of art and technology" to give artistic direction to industry, which was as lacking in 1919 as in the mid-19th century, when the Arts and Crafts movement began. The greatness of Gropius as an educator was that he did not put forward any dogmatic policies, but rather he acted as a balance between the rational, representative, and physical on the one hand and the spiritual, esthetic, and humanitarian on the other. An artistic community of prima donnas is difficult to coordinate, but Gropius acted as choreographer and exacted the best from his faculty, from the mysticism of Johannes Itten to the Marxist socialism of Hannes Meyer.

When right-wing criticism forced the Bauhaus to leave Weimar in 1925, Gropius designed the structure for the new Bauhaus in Dessau, one of his finest works, which embodied a new concept of architectural space. When criticism mounted there against him as director in 1928, he resigned rather than allow the criticism to spread from him as leader to the whole institution. (Nazism and the Bauhaus stood for diametrically opposing viewpoints, and in 1933 under Ludwig Mies van der Rohe the school, which had moved to Berlin, was forced to close.)

Gropius practiced in Berlin from 1928 to 1934, experimenting with prefabricated housing in his Toerten housing development in Dessau (1926) and dwellings at the Werkbund Exhibition (1927). He went to England in 1934, where he worked with E. Maxwell Fry until 1937, designing mainly individual houses, but also Impington College, Cambridgeshire. This structure partially influenced the post-World War II school design program in Britain.

Works in America

When Gropius went to the United States in 1937, he collaborated with Marcel Breuer, a former pupil, on individual and group housing, including a house for himself at Lincoln, Mass. (1937). Gropius held the chair of architecture at Harvard from 1938 to 1952, a period of his life from the age of 55 to 69, when most architects would have been designing their major works. This was due to his intense commitment to the educational process. "I have been 'nobody's baby' during just those years of middle life which normally bring a man to the apex of his career," Gropius admitted, when he received the American Institute of Architects' Gold Medal in 1959.

Gropius had, however, established The Architects' Collaborative (TAC), a group-oriented practice, in 1946, and he retired from Harvard in 1952 to devote his full attention to the practice of architecture. TAC and Gropius designed the Harvard University Graduate Center (1949—1950); executed a project for the Boston Back Bay Center (1953), which was not carried out; and designed the U. S. Embassy in Athens (1960) and Baghdad University in Iraq (begun 1962 but incomplete as of 1971).

Gropius also designed locomotives and railroad sleeping cars (1913—1914), the Adler automobile (1930), and a host of everyday products. He believed in "the common citizenship of all creative work."

Free Reading 2

The Ulm School of Design

The Ulm School of Design (Hochschule für Gestaltung—HfG Ulm) existed for 15 years in Germany from 1953 to 1968. It was founded by Inge Aicher-Scholl, Otl Aicher and Max Bill, who was a student at the Bauhaus. HfG Ulm is considered to be the most influential design school in the world after the Bauhaus. Although the school closed in 1968, the "Ulm Model" concept continues to influence international design education. Notable teachers and students include Tomás Maldonado, Walter Zeischegg, Peter Seitz, and John Lottes.

The Ulm School of Design was one of the most progressive institutions for teaching design and environmental design in the 1950s and 1960s. It was founded by Inge Scholl, Otl Aicher, and Max Bill; the latter became the school's first rector in 1953. The HfG Ulm rapidly gained international recognition. New approaches in design were investigated and put into practice within the departments: Visual Communication, Industrial Design, Building, Information, and later Film. The HfG building was designed by Max Bill and is still an impressive sight today, with its prominent open setting.

The campus reflects the teaching concept, namely the integration of work and life in one place. The history of the HFG was shaped by innovation and change, consistent with the school's own image of itself as an experimental institution. This resulted in countless modifications to the content and organization of classes.

Graphic Works

350 objects and prototypes from classes in the Industrial Design and Building Departments, serially-produced products from HfG development group designs. The objects were designed to be suitable for daily use, serial production, and industrialized construction (See Figure 2.12 and Figure 2.13). The objects in the collection were created in industrial design classes taught by Bill, Bonsiepe, Gugelot, Leowald, Lindinger, Maldonado, and Zeischegg. Those from the Building Department come from classes taught by Bill, Doernach, Ohl, Schnaidt, and Wirsing.

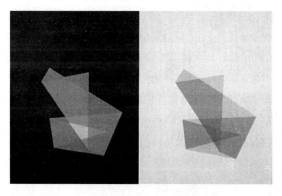

Figure 2.12 Transparency, academic year 1953—1954
Lecturer: Josef Albers
Student: Ingela Albers

Figure 2.13 Control Unit Studio 1, 1957
Hans Gugelot and Herbert Lindinger

11,000 black and white negatives with contact prints, around 3,000 black and white original enlargements and 6,000 slides, shots of the HfG building, student work, serial products, photos taken of classes and events, portraits of lecturers, taken with the primary aim of documenting and presenting the school are shown in Figure 2.14 and Figure 2.15.

Exhibition Panels

350 exhibition panels compiled at the school about the school and work produced in classes, which include, for example, the 1958 exhibition to mark the fifth anniversary of the HFG's founding, shown in Figure 2.16, the photography exhibition of 1963—1964, and landscape and product photos.

Figure 2.14 Aerial shot of the HfG building

Figure 2.15 Student meeting, 1955

Figure 2.16 HfG exhibition in the refectory and auditorium, 1958

Free Reading 3

Transformation Paves the Way for Furture Designers——Design Education at Polytechnic University

Transformation is the keyword for the new BA (Hons) programme of the polytechnic University's School of Design. The new curriculum is a logic follow-through of the University' credit-based system which signifies transition from a technical institute to university, While the old curriculum channeled students into different disciplines such as graphics, interior, industrial, photography and fashion design, the new programme unites them with the aim of equipping design students with flexibility and versatility.

Anticipating the need for designers in the 21th century to have wide-ranging knowledge and muti-and inter-disciplinary understanding, as well as business and management skills, the School of Design intiated this new programme of design education starting in September 1998. The Head of the School of Design, Prof. John H. Frazer said, "Things are going to change very rapidly in the design field in the next few years and the old style of teaching didn't have the flexibility to respond to changing needs... We feel it's our obligation to train people as far as we can, to encourage flexibility in particular in a fast-paced world so that our designers are prepared when they go out and practice."

Interdisciplinary Learning and Teaching

Interdisciplinary learning and teaching are the focus of the new programme. Students are able to work co-operatively with peer groups under the new curriculum. "The change reflects the philosophy of interdisciplinary learning and teaching. In fact, the School is making fundamental changes because we want to support the curriculum with actual facilities," said Alice Lo, the Programme Leader.

Walls were knocked down to make space for the new studios. The re-planning has made the school more spacious and, most importantly, prompted students to exchange ideas among themselves. Without walls, students are doing a lot of peer group learning. The "classroom" concept does not exist in the new studio-students' lockers can be moved around and different areas are separated by movable partitions. This way, the place can be rearranged to become a lecture hall, a form for small group discussion, or a work stations for individual study. It becomes a labyrinth of movement encouraging flexibility and the exchange of ideas.

Sitting next to each other will be students from different disciplines but who are working on the same project. Thematic projects are designed to help eliminate the boundaries between different disciplines. The teaching team is composed of tutors and instructors from different disciplines, guiding students to approach the projects from their own viewpoint. Prof. Frazer, when describing the thematic projects specially designed for the students, said, "All the different course students have agreed that there are commen themes. Even when (the students) trained to become specialists, there is a commonality whereby they can talk to each other and work jointly towards the end result." Peer group assessment will also be encouraged so that students benefit from understanding the assets as well as limitations of various disciplines apart from their own.

A special feature of the new curriculum is that there will be a virtual studio, in which a supercomputer will be installed. It will be a paradoxical kind of space. When one is inside this confined chamber—what Prof. Frazer calls a "black box"—and puts on special glasses, she is actually connected, not only to the people around, but also to the rest of the world. With such hi-tech facilities, students will be able to do join hands with their counterparts with students from other parts of the world.

Some Reflections

While being enthusiastic about the birth of a new program, there are risks in interdisciplinary. Programes and the arduous effort it entails to be successful. The students who are enrolled in such a programme should have the ability to apprehend different aspects of various disciplines while remaining firmly rooted in their own. One can only compare different discilines or conduct interdisciplinary studies when he/she has in-depth knowledge of his/her own discipline.

Students now are gifted with hi-tech facilities. They should be versed in the computer system and how best to us it. They should also bear in mind that using the facilities an end in itself: the facilities are only tools aiding them to create better designs.

The new curriculum in the School of Design is an expression of the desire to improve. As Prof. Frazer said, things are going to change rapidly in the design field. The new programme does not aim at training traditional designers but "designers who can invent new future."

Free Reading 4

Braun

This German electrical and audio-visual equipment manufacturer has for more than half a century been associated with high-quality designs, many of which have featured prominently in museum collections and design competitions around the world. Its origins lay in the radio accessory manufacturing company founded in 1921 by Max Braun near Frankfurt. However, it was not until after the end of the Second World War that the company expanded its product range to include the domestic appliances upon which its international reputation soon began to emerge.

Following the death of Max Braun in 1951 his sons Artur and Erwin took over the management of the company, diversifying into electric razors and audio equipment, including the *Kombi* radiogram by Wilhelm Wagenfeld. However, the establishment of the *Hochschule für Gestaltung* (HfG) at Ulm in 1953 had a profound influence on the design of Braun products, cemented by the appointment of Dr. Fritz Eicher, a lecturer from the HfG, as head of Braun's design department in 1956.

The clean Modernist appearance of Braun products was closely identified with the international "Good Design" ethos of the 1950s and 1960s and was in complete contrast to the extravagant styling of many ephemeral products, an iconic potential that was confirmed by the New York Museum of Modern Art's exhibition of Braun products in 1964 (a number of which had featured in MOMA's permanent collection since 1958). However, the company's absolute and exacting commitment to quality design did not equate with commercial success and the company was taken over by Gillette in 1967. This facilitated greater international market penetration and product diversity across a number of fields, from hairdryers to coffee machines and calculators to electric toothbrushes. Nonetheless the functionalist appearance remained an essential ingredient of the Braun agenda. By the 1990s, with the consumer appetite for originality and wit in everyday products undiminished by the economic recession of the 1980s, many Braun products took on a more colourful, less restrained appearance in order to remain competitive in the highly competitive market place for domestic goods.

From 1984 through 2005, Braun was awholly owned subsidiary of the Gillette Company, which had purchased a controlling interest in the company in 1967. Braun is now a wholly owned subsidiary of Procter & Gamble, which acquired Gillette in 2005.

Products

Braun's products include the following categories:
- Shaving and Grooming (electric shaving, hair trimming, beard trimming)
- Oral Care (now under the Oral-B brand)
- Beauty Care (hair care and epilation)
- Health and Wellness (ear thermometers, blood pressure monitors)
- Food and Drink (coffee makers, coffee grinders, toasters, blenders, juicers)
- Irons
- Clocks and Calculators

History

Max Braun, a mechanical engineer, established a small engineering shop in Frankfurt am Main in 1921. In 1923 he began producing components for radio sets. In 1928 the company had grown to such an extent, partly due to the use of certain plastic materials that it moved to new premises on *Idsteiner Strasse*.

Eight years after he started his shop, Max Braun began to manufacture entire radio sets in 1929. Soon after, Braun became one of Germany's leading radio manufacturers.

In 1935 the Braun brand was introduced, and the familiar logotype with the raised "A" took form. At the 1937 World's Fair in Paris, Max Braun received the award for special achievements in phonography. Three years later, the company had more than 1000 employees.

Braun continued to produce state-of-the art radios and audio equipment, and in 1956 introduced its now famous SK-4 record player. Braun soon became well known for its "high-fidelity" audio and record players.

The 1950s also marked the beginning of the product that Braun is most known for today: the electric shaver. The S50 was the first electric shaver from Braun. The shaver was designed in 1938, but World War II delayed its introduction until 1951. It featured an oscillating cutter block with a very thin, yet very stable steel-foil mounted above it. This principle is still used in Braun's shavers of today. Some excellent works are as follows: Figure 2.17—Figure 2.20.

Figure 2.17 1960 headquarter of the company in Frankfurt am Main

Figure 2.18 Braun Sixtant SM2

Figure 2.19 Braun SK5, nicknamed Snow White's coffin

Figure 2.20 Braun HF 1, Germany 1958

Core Text 7

Raymond Loewy

Raymond Fernand Loewy (5 November 1893—14 July 1986, as shown in Figure 2.21) was one of the best known industrial designers of the 20th century. Born in France, he spent most of his professional career in the United States where he influenced countless aspects of North American culture. His career spanned seven decades. To a degree unequaled by the names of any of the other founding fathers of industrial design, the name of Raymond Loewy radiates a charisma that has attracted public attention throughout the past half century. Loewy's design philosophy is not a deeply intellectual one. He summarized it with the acronym MAYA (most advanced, yet acceptable).

He studied (1918) for a degree in engineering at the Ecole de Lanneau in Paris, before serving in the French Army. In 1919 he emigrated to the USA (naturalized 1938). After a brief period where he worked as a fashion illustrator and designer of department store window displays. Working in advertising design steered him towards industrial design. His disappointment with the quality and vulgarity of American products led him in 1929 to design and re-style radios for Westinghouse and duplicating machines for Gestetner. In 1934 he designed the cold spot refrigerator for Sears Roebuck. Loewy's success was based on his rehousing of American products in streamlined forms in the 1930s. From the 1940s to the 1970s he worked on a number of commissions, such as the corporate logos of many major businesses, including British Petroleum and Shell Oil.

Other successful projects were the designs for the S-I locomotive (shown in Figure 2.22) for the Pennsylvania Railroad Company, styling for Studebaker cars (including the Avanti) from 1937 to 1962, the Dole Coca Cola dispenser (c. 1948), architectural designs for International Harvester dealer showrooms and interiors of NASA's Saturn-Apollo and Skylab projects (1967—1973). He had offices in many American and foreign cities and enjoyed great success until the 1980s, when financial pressure led to the contraction of his business.

Figure 2.21 Raymond Loewy

Figure 2.22 Raymond Loewy standing on one of his designs, the Pennsylvania Railroad's S1 steam locomotive

It is difficult to measure precisely Loewy's impact on our contemporary environment, but he has certainly had a dynamic and significant one. His continuing vitality and international influence are demonstrated by his being retained as a major industrial design consultant in the 1970's by the Government of the Soviet Union—this as he entered his eighties.

Avanti

In the spring of 1961, Loewy was called back to Studebaker by the company's new president, Sherwood Egbert, to design the Avanti. Egbert hired him to help energize Studebaker's soon-to-be-released line of 1963 passenger cars to attract younger buyers. (shown in Figure 2.23 and Figure 2.24)

Figure 2.23 Concept sketch of the 1963 Avanti by Loewy

Figure 2.24 1963 Studebaker Avanti

Despite the short 40-day schedule allowed to produce a finished design and scale model, Loewy agreed to take the job. He recruited a team consisting of experienced designers including former Loewy employees John Ebstein, Bob Andrews, and Tom Kellogg, a young student from Art Center College of Design in Pasadena. The team was sequestered in a house leased for the purpose in Palm Springs, California. Each team member had a role: Andrews and Kellogg handled sketching, Ebstein oversaw the project, and Loewy was the creative director and offered advice.

The Avanti became an instant classic when it hit the market and has many devotees today; others consider its front end styling peculiar. Versions have been produced in limited quantities over the years by a succession of small independent companies, though never with real commercial success.

Death and Legacy

Loewy became a U.S. citizen in 1938. He died in 1986 at the age of 93. In 1991, the Raymond Loewy Foundation was established in Germany to further promote the discipline of design internationally and preserve the memory of Raymond Loewy. An annual award of 50,000 Euros is granted to outstanding designers in recognition of their lifetime achievements. Grantees have included Phillippe Starck and Dieter Rams. He was married twice.

His marriage to Jean Thomson ended in amicable divorce in 1945. He married Viola Erickson in 1948. Their daughter, Laurence, managed her father's interests in the United States after his death. She died October 15, 2008, in Marietta, Georgia.

Key Words

[1] industrial designer *n.* 工业设计师

[2] unequaled [ˌʌn'iːkwəld] *adj.* 无与伦比的；无双的；不等同的

[3] charisma [kə'rɪzmə] *n.* 1. 魅力，魔力 2. 〈宗〉神授的力量或才能 3. 超凡的个人魅力；感召力；号召力

[4] Avanti-design *n.* 阿凡提设计

[5] contemporary [kən'tempərərɪ] *adj.* 1. 当代的；现代的 2. 同时代的，同属一个时期的

[6] advertising design 广告设计

[7] vulgarity [vʌl'gærɪtɪ] *n.* 1. 庸俗，粗俗，粗鄙 2. 粗野的行为；粗俗的话

[8] architectural design 建筑设计

[9] devotee [ˌdevə'tiː] *n.* 热爱者

[10] grantee [ɡrɑːn'tiː] *n.* 受让人

Key Sentences

1. To a degree unequaled by the names of any of the other founding fathers of industrial design, the name of Raymond Loewy radiate a charisma that has attracted public attention throughout the past half century. Loewy's design philosophy is not a deeply intellectual one. He summarized it with the acronym MAYA (most advanced, yet acceptable).

作为工业设计之父，雷蒙德·罗维是无人能及的。雷蒙德·罗维这个名字散发的魅力曾经吸引公众注意长达半个世纪之久。罗维的设计理念并非深不可测，他将其总结为"MAYA"(Most Advanced, Yet Acceptable 最先进的，尚可接受的)。

2. It is difficult to measure precisely Loewy's impact on our contemporary environment, but he has certainly had a dynamic and significant one.

准确衡量罗维对现代环境的影响并非易事，但他肯定是充满活力并且具有重要意义的。

课文翻译

雷蒙德·罗维

雷蒙德·罗维(1893年11月5日—1986年6月14日)是20世纪最著名的工业设计师之一。他生于法国，但职业生涯都在美国，在那里，他影响了北美文化的诸多领域。他的职业生涯持续了70年。工业设计之父雷蒙德·罗维是无人能及的。雷蒙德·罗维这个名字散发的魅力曾经吸引公众注意长达半个世纪之久。罗维的设计理念并非深不可测，他将其总结为"MAYA"(Most Advanced, Yet Acceptable, 最先进的，尚可接

受的)。

在法国服役之前,他在巴黎的格勒诺布尔管理学院攻读工程学位(1918)。1919年,他移居到美国(1938年加入美国国籍)。之后一段很短的时期,他是百货公司的橱窗展示设计师和时尚插图画家。广告设计工作推进了他向工业设计发展。他对美国产品的质量和庸俗失望,这导致了他在1929年为威斯汀·豪斯公司设计,并且重新改造收音机的造型,为基士得耶公司重新设计了复印机。1934年,他为西尔斯·罗巴克公司设计了冷点冰箱(cold spot refrigerator)。罗维的成功基于他在20世纪30年代为美国产品重新制定了流线型生产线。从20世纪40年代到70年代,他得到大量的企业委托,例如,设计许多大企业的企业标志,包括英国石油公司和壳牌石油公司。

其他成功案例包括1937年为美国宾西法尼亚州铁路公司设计S-I火车机车,1937年到1962年间为Studebaker汽车设计新样式,1948年可口可乐的瓶子。在建筑设计上,为国际收割机商人设计展室和1967年至1973年间为美国国家航空航天局的"Saturn-Apollo"设计舱室和太空实验室项目。他在美国和许多其他国家的城市都设有办公室,获得了巨大成功。直到20世纪80年代,经济压力导致他业务缩减。

准确衡量罗维对我们现代环境的影响并非易事,但他肯定是充满活力且具有重要意义的。20世纪70年代,当他在80岁的时候,他被苏联政府聘为重要的工业设计顾问,这足以证明他持续的生命力和国际影响力。

阿凡提

1961年春天,罗维被公司新总裁舍伍德·埃格伯特召回到史蒂倍克(Studebakeg)公司设计阿凡提(Avanti)。埃格伯特雇他来帮助史蒂倍克公司即将推出的1963年汽车生产线,以吸引年轻买家。

虽然计划表上完成设计到制作模型只有40天,但是他还是同意接受这项工作。他招募了一支由经验丰富的设计师们组成的团队,包括罗维的前雇员约翰·艾博斯坦、鲍勃·安德鲁以及一名来自于巴萨迪那艺术中心设计学院的年轻学生汤姆·克劳格。这个团队隐居在加利福尼亚州棕榈泉的一所出租房里,每个成员都有一个角色:安德鲁和克劳格负责草稿,艾博斯坦监督整个项目,罗维作为创意总监并提供建议。

阿凡提在进入市场的瞬间就成为经典,直到今天仍有许多爱好者。反对者认为它的前端造型奇特。多年来,它的版本已经被一些小型的个体公司生产为限量版,虽然它从来没有真正获得商业上的成功。

去世和遗产

1938年,罗维成为美国公民。他于1986年去世,享年93岁。1991年,为了激励国际设计学科并纪念罗维,德国设立了罗维基金。奖金每年5万欧元,用来奖励那些被认为在有生之年获有成就的杰出设计师。受捐助人包括菲利普·斯塔克和迪特尔·拉姆斯。罗维结过两次婚,他与珍·汤姆森的婚姻在1945年友好结束。1948年,他与维奥拉·埃里克森结婚。在他去世后,他们的女儿劳伦斯在美国管理父亲的资产。她于2008年10月15日在佐治亚州玛丽埃塔去世。

Free Reading 1

Streamline Modern

Definition

Streamline Modern, sometimes referred to by either name alone, was a late branch of the Art Deco style. The style emphasized flowing forms, long horizontal lines, and sometimes nautical elements (such as railings and porthole windows). The buildings in Frank Capra's 1937 movie *Lost Horizon*, designed by Stephen Goosson, exemplify the soothing style.

Profile

From the early 1930s through into the 1950s, in the United States a design style flourished that has become known as the Streamline Style. Its most important characteristics are the closed, streamlined forms that strongly suggest speed, symbolic of the dynamism of modern times. This style dangled the promise before consumers, racked by the economic crisis, that they were still on the way to a glorious future with prosperity for everybody, at least if they continued to consume.

The Streamline Style stood for mobility, speed, efficiency, luxury and hygiene, all concepts that were identified with modernity. To visualise this, the sharp corners and transitions of objects were rounded off.

Norman Bel Geddes

Norman Bel Geddes (1893—1958) was atheatrical set designer who turned his hand to applying Streamline Style to vehicles, such as fantastic (and non-airworthy) aeroplanes, space-age cars and super-stylised trains. He designed the famous General Motors Pavilion for the 1939 New York World Fair, which included the Highway and Horizons exhibit, more commonly known as "Futurama". Few of his vehicle designs were ever actually made, but his imagination and sheer style captured the public's attention.

Notable Examples of Streamline Moderne:
- 1933—Merle Norman Building, Santa Monica, California
- 1934—Chrysler Air-Flow, the first mass-market streamline automotive design
- 1935—Pan Pacific Auditorium, Los Angeles, California
- 1935—The Hindenburg, zeppelin passenger accomodations
- 1937—Belgium Pavilion, at the Exposition Internationale, Paris
- 1937—TAV Studios (Brenemen's Restaurant), Hollywood, California
- 1937—Minerva (or Metro) Theatre and the Minerva Building, Potts Point, New South Wales
- 1937—Barnum Hall (High School auditorium), Santa Monica, California
- 1939—Marine Air terminal, La Guardia Airport, New York

Some representative works of Streamline Style are as follows: Figure 2.25 Buick LeSabre (1951)—Harley Ea, Figure 2.26 Fada 1000L Bullet radio (1946) image sourced here.

Figure 2.25 Buick LeSabre (1951) —Harley Ea

Figure 2.26 Fada 1000L Bullet radio (1946) image sourced here.

Free Reading 2

Ergonomics

In August 2000, IEA Council adopted an official definition of ergonomics as following.

The Discipline of Ergonomics

Ergonomics (or human factors) is the scientific discipline concerned with the understanding of interactions among humans and other elements of a system, and the profession that applies theory, principles, data and methods to design in order to optimize human well-being and overall system performance.

Ergonomics contribute to the design and evaluation of tasks, jobs, products, environments and systems in order to make them compatible with the needs, abilities and limitations of people.

Domains of Specialization

Derived from the Greek *ergon* (work) and *nomos* (laws) to denote the science of work, ergonomics is a systems-oriented discipline which now extends across all aspects of human activity. Practising ergonomists must have a broad understanding of the full scope of the discipline. That is, ergonomics promotes a holistic approach in which considerations of physical, cognitive, social, organizational, environmental and other relevant factors are taken into account. Ergonomists often work in particular economic sectors or application domains. Application domains are not mutually exclusive and they evolve constantly; new ones are created and old ones take on new perspectives.

There exist domains of specialization within the discipline, which represent deeper competencies in specific human attributes or characteristics of human interaction.

Domains of specialization within the discipline of ergonomics are broadly the following:

Physiological and biomechanical is concerned with human anatomical, anthropometry, physiological and biomechanical characteristics as they relate to physical activity. (Relevant topics include working postures, materials handling, repetitive movements, work related musculoskeletal disorders, workplace layout, safety and health.)

Cognitive ergonomics is concerned with mental processes, such as perception, memory, reasoning, and motor response, as they affect interactions among humans and other elements of a system. (Relevant topics include mental workload, decision-making, skilled performance, human-computer interaction, human reliability, work stress and training as these may relate to human-system design.)

Organizational ergonomics is concerned with the optimization of socio-technical systems, including their organizational structures, policies, and processes.

Ergonomics Applying: Herman Awarding Furniture Design

With the Resolve system, 120-degree angles form open, inviting workstations where people feel comfortable, welcome, and connected. That's especially important for those who collaborate in a community of work. For computer users, Resolve supports complex technology in a corner far more generous than 90 degrees. The 120-degree geometry also allows more planning options and efficient use of materials.

Resolve—A Pole-based System

In the Resolve pole-based system, vertical poles, in conjunction with horizontal support arms, define the work space, provide a structural foundation, support hang-on components, and make power and data accessible to the user. Trusses attach to tall pole tops to route power and data cables overhead.

120-Degree Geometry

Optimum floor plan: Uses space economically and efficiently. Inviting angle: Creates open and welcoming workstations where people can collaborate effectively. Creative design possibilities: Provides more planning options, layout flexibility, and many levels of enclosure.

Personalization

Rolling screens: User control privacy and enclosure.

Canopies and floor mats: define personal territory.

Customized graphics: design on textile allows digital imaging on screens, canopies, and flags.

Work tools: People can keep them right where they're needed.

Human scale: The design is based on the size, reach, and movements of people.

Technology Support

Overhead pathway: Trusses carry power and data independent of the work space below, so configuration changes don't affect delivery.

Easy installation: Cables are laid into trusses, which hold 100 Category 5, 4-pair UTP cables.

Generous power: 6-circuit, 10-wire power allows circuit configurations of 4 general and 2 isolated, 3 general and 3 isolated, or 2 general and 4 isolated.

Convenient access: Each pole has locations for 9 duplex receptacles.

Doing More with Less

Efficiency, integrity: Resolve uses less material to do more, delivering high performance affordable with environmental responsibility. Concise vocabulary: With about 1/4 the number of products of most systems, Resolve helps simplify planning, specifying, ordering, and installation. Simple, lightweight infrastructure: Installation and reconfiguration are faster. Translucent boundaries: Light-scaled screen define territories while keeping people accessible and connected.

Design Story

Herman Miller and designer Ayse Birsel saw the sweeping and ongoing changes in people, the rise of collaborative work, and the expansion of complex technology as a great opportunity to re-examine the work environment and resolve critical issues.

The result of this inquiry is the Resolve system. Resolve isn't an evolution of panel-based products all the intelligence and function of panels while allowing greater diversity of workstation patterns, as well as more efficient use of materials and real estate. There's also more openness and flexibility for collaborative environments.

Resolve's human scale and ease of use reflect its fundamental design principle: It all starts with the person, giving each worker the tools and capabilities to be more comfortable, connected, and effective.

Free Reading 3

Corporate Identity

The management and communication of a company's identity increasingly is being viewed by senior executives as vital to corporate success. The concept of corporate identity can be traced to the earliest firms that used specific marks or logos to differentiate themselves from their competitors and imprint their image in the minds of consumers. By the 1970s a robust consulting industry that specialized in helping companies improve their image had emerged. More recently, in response to the dynamics of the business environment, many of these designconsultants have broadened their focus to embrace a strategic view of communicating corporate identity. With this amplified focus, the epithet for the concept evolved from "corporate image" to "corporate identity" (and, more and more, the term "corporate brand" is being applied).

Importance of Corporate Image

The growing significance of managing corporate identity is underscored by a 1989 survey in Britain by Market Opinion Research International, which found that 77 percent of the leading industrialists questioned believed that the importance their firms attached to developing and promoting their corporate identity would increase in the near future. Research a year later by CBI and Fitch Consultants corroborated this finding and the experience of the 1990s strongly suggests that this expectation has materialized.

The overriding reason for the burgeoning concern for corporate identity is abundantly clear. We live in a time of immense environmental complexity and change, and consequently corporations have been forced to significantly alter their strategies to better compete and survive. Mergers, acquisitions, and divestitures represent a major dimension of corporate change over the past several decades. Consider the extreme example of the Greyhound Corporation. For most of this century, Greyhound was the largest busing company in North America. In the 1970s, however, the company initiated an aggressive acquisition/diversification and by the late 1980s was competing in five different industries (it even sold off most of its busing operations). To signal this metamorphosis to its external audiences, the company belatedly changed its name to the Dial Corporation and completely revamped its corporate communications.

The acceleration of product life cycles is another vital dimension of the turbulent business environment. Nowhere is this more apparent than in the electronics industry. Personal computers can become outmoded in the period of less than a year. In the audio segment of the market, tapes replaced records and, in turn, were replaced by compact discs, which may in the future be superseded by digital audiotapes. Companies with strong corporate images, such as Sony Corporation and Casio, obviously have an advantage in such dynamic markets because their name adds value to their products by reducing uncertainty in the eyes of distributors, retailers, and consumers.

Deregulation has been a critical factor in many industries. For instance, as a result of the court-ordered breakup, AT&T has had to develop a new strategy and a more aggressive marketing-oriented culture to adjust to its new realities. Concurrently, the telecommunications giant adopted a new logo and initiated a communication program to help convey its new identity. In another example, deregulation of the financial services industry has allowed savings and loan associations to expand their services and compete with banks. Consequently, firms such as Glendale Federal Savings and Loan Association and California Federal Savings and Loan Association have changed their charters to become banks. Their new names are Glendale Federal Bank and California Federal Bank, respectively, and they have fittingly redirected their corporate communication programs.

Globalization has been still another catalyst in the rise of corporate identity programs. To illustrate, American Express Co. originally was a freight company in the North American market. As the company matured into a global credit card, banking, and travel organization, it wisely developed a corporate communication program aimed at projecting its new identity. American Express understood that a strong and positive global image can be a powerful weapon for firms expanding internationally. IBM, McDonald's, and Baskin-Robbins are examples of other companies that have been able to expand to all areas of the world with relative ease because of their global prominence.

A related factor is that as a corporation expands its operation internationally, or even domestically, through acquisitions, there is a danger that its geographically dispersed business units will project dissimilar or contrary images to the detriment of corporate synergy.

British-based Courtaulds has a globally dispersed organization but until its latest identity review allowed its operating companies to use their traditional names. As a consequence of this policy, there was little cooperation among these units and no cohesive corporate identity. Courtaulds remedied this problem by instituting a common naming policy and a correlated corporate communication program.

Still another factor stimulating the current interest in corporate identity is society's growing expectation that corporations be socially responsive. One salient manifestation of this trend is that many of today's consumers consider the environmental and social image of firms in making their purchasing decisions. Companies such as Ben and Jerry's and Tom's of Maine have built their strategies around this idea and consequently have grown very rapidly. Another manifestation of the trend is the rise of socially responsible investment funds.

Theory of Corporate Identity

Theory always underlies good practice. Theory identifies and defines the key variables in the process under consideration and explains the interrelationship among them. In the process for managing corporate identity, the fundamental variables are corporate identity, corporate communication, corporate image, and corporate reputation. Corporate identity is the reality of the corporation. It is the unique, individual personality of the company that differentiates it from other companies. To use the marketing metaphor, it is the corporate brand. Corporate communication is the aggregate of sources, messages, and media by which the corporation conveys its uniqueness or brand to its various audiences. Corporate image and corporate reputation are in the eye of the beholder. Image is the mental picture that people have of an organization, whereas reputation constitutes a value judgment about the company's attributes.

The interrelationship among these variables is shown diagrammatically in Figure 2.27. The objective in managing corporate identity is to communicate the company's identity to those audiences or constituencies that are important to the firm in a manner that is both positive and accurate. This process involves fashioning a positive identity and communicating this identity to significant audiences in such a way that they have a favorable view of the company. The feedback loops in the model indicate that an unsatisfactory image or reputation can be improved by modifying corporate communication or reshaping the corporate identity or both. The principal issues relating to the five components of the model—identity, image, reputation, communication, and feedback—will now be examined in greater detail.

Components of Corporate Identity

Corporate identity, as explained above, is the reality and uniqueness of the organization. It may be broken down into its component parts: corporate strategy, corporate culture, organizational design, and operations. Strategy is the overall plan that circumscribes the company's product/market scope and the policies and programs by which it chooses to compete in its chosen markets. For example, Southwest Airlines is a regional carrier competing in the airline industry through strategies that result in low costs and low fares.

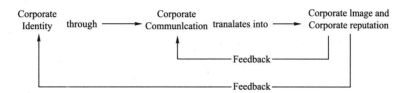

Figure 2.27 Corporate Image Model

Corporate culture is the shared values, beliefs, and assumptions that the organization's members hold in common as they relate to each other, their jobs, and the organization. It defines what the firm personnel believes is important and unimportant, and explains to a large degree why the organization behaves the way it does. Southwest Airlines has a strong corporate culture that highly prizes company loyalty, internal cooperation, and service to the customer. Southwest's culture supports the company's strategy and is a prominent component of its identity.

Organizational design refers to the basic choices top managers have in developing the pattern of organizational relationships. It encompasses issues such as whether basic departmentation should be by function or product division, the overall configuration (tall vs. flat), the degree of decentralization, the number of staff personnel, the design of jobs, and the internal systems and procedures. All of these factors can affect, to some degree, corporate identity. From the perspective of the firm's external constituents, however, the corporate/product relationship normally is the most critical element of organizational design. The corporate/product relationship refers to the deliberate approach a firm follows in structuring the relationship of its products to one another and to the corporate entity. Corporate/product relationships may be categorized as single entity, brand dominance, equal dominance, mixed dominance, or corporate dominance.

Single entity companies offer one product line or set of services; consequently, the image of the company and that of the product tend to be one and the same. Southwest Airlines is an obvious example of a single entity company; it is 100 percent involved in the airline business. Identity problems typically arise for single entity firms such as Southwest Airlines if they expand into areas and activities not immediately related to their current strategy. The corporate planners must carefully consider the corporate identity they desire to have and the concomitant image they wish to project.

Under the brand-dominant approach, the decision has been made not to relate the product brand and corporate names. This approach is followed by many consumer products companies. For instance, Marlboro and Merit cigarettes, Post cereals, Jell-O, Kraft cheeses, and Oscar Mayer meats are all well-known products but few consumers realize that they are all marketed by the Philip Morris Companies, Inc.

General Motors Corp., historically, has exemplified the equal-dominance approach. The principal General Motors' automobile divisions—Chevrolet, Pontiac, Oldsmobile, Buick, and Cadillac—maintained separate identities, but each was also closely associated with the corporation. Neither the corporate nor the individual brand was predominant.

In mixed-dominance companies, sometimes the brand name is dominant, sometimes the corporate name is dominant, and in some cases they are used together with equal emphasis. The German firm Robert Bosch GmbH follows this approach. Bosch identifies some of the products it manufactures (for example, spark plugs) with the corporate name, but chooses to allow other brands such as Blaupunkt radios to stand independently.

IBM, Hewlett-Packard, Xerox Corporation, and Gerber Products Co. are exemplars of the corporate-dominance strategy. In these companies, the corporate identity is paramount and all implementation decisions are aimed at reinforcing this identity.

Two examples will illustrate the identity considerations involved in corporate/product relationship decisions. A number of years ago, Pillsbury developed a franchised restaurant chain called Bennigan's Tavern. The company had to decide if it wanted to create an identity for Bennigan's that linked the restaurant to Pillsbury or create a separate identity for it. A crucial question was whether a corporation known for its traditional kitchen products such as flour and cake mixes should associate its name with an institution serving beer. Prudently, Pillsbury chose the brand dominant approach.

The classic example of the Transamerica Corporation illustrates an equally logical but very different outcome. Transamerica was a conglomerate with divisions in such diverse fields as air travel, business forms, entertainment, and insurance. For many years the company followed a brand dominant (or more precisely, a subsidiary-dominant) approach in which the relationships between the subsidiaries and the parent were deemphasized. In the late 1960s, however, company executives decided to take advantage of the synergies that a unified corporate identity could render. As part of the implementation program, Transamerica created the "T" logo as a unifying symbol to connect the various subsidiaries to the parent and to each other. Each subsidiary retained the company name under which its reputation had been built, and the corporate relationship was communicated through a linking phrase such as, "Entertainment from the Transamerica Corporation".

Operations, the fourth and final component of corporate identity, are the aggregate of activities the firm engages in to effect its strategy. These activities become part of the reality of the corporation and can influence its image and/or reputation in a wide variety of ways. Several examples will highlight the range of possibilities. First, the cleanliness of Disney Corporation's theme parks, together with the efficiency and helpfulness of their employees, has become a significant and very positive dimension of the corporation's identity. At the other end of the spectrum, the inappropriate treatment of African-American customers at some Denny's restaurants negatively affected that company's overall reputation as did the now infamous *Exxon Valdez* oil spill for the Exxon Corporation.

Corporate Image and Reputation

Corporate image and reputation are discrete but related concepts. As noted earlier, corporate image is the effigy that people have of a company. Corporate reputation, on the other hand, represents a value judgment that people make about the firm as a whole or one or more of its attributes. Corporate images typically can be fashioned fairly quickly through specific actions and well-conceived communication programs, whereas reputations evolve over time as a result of consistent performance (and they can be reinforced through corporate communication).

Clearly, a corporation must be concerned about its image and reputation amongst its important constituent groups. In academic parlance, these significant constituent groups are called stakeholders. They are groups that have a stake in the company. Stakeholders are affected by the actions of the company and, perhaps more importantly, their actions can affect the company. Consequently, its image and reputation in the eyes of its stakeholders is critical to the company. The principal stakeholders with which most large firms must be concerned are:

- Customers
- Distributors and retailers
- Financial institutions and analysts
- Shareholders
- Government regulatory agencies
- Social action organizations
- The general public
- Employees

The company's image and reputation vis-a-vis its various stakeholders will influence their willingness to provide or withhold support. Thus, if its customers develop a negative perception of the company or its products, its sales and profits assuredly will decline. Consider the recent travails of the Nissan Motor Company. In the 1980s it enjoyed the image of a customer-oriented, trendsetting automobile manufacturer with an excellent reputation for automotive engineering. By the mid-1990s, however, as a result of a series of poor decisions, its image as a cutting-edge producer, along with sales and profits, had declined precipitously. It is now perceived by customers as well as other stakeholders as a conservative maker of stodgy, boxy cars with its engineering reputation compromised.

The impact of corporate identity in the financial community can be seen through the history of the British packaging, printing, and coating company that recently changed its name from Bowater to Rexham in response to confusion in the financial community as well as among its customers as to its identity. In North America the company traded under the name Rexham, whereas in the rest of its markets it operated under the Bowater banner. The name change was initiated by itschief executive officerto create the image of a global competitor in the eyes of financial institutions and investors, as well as its customers.

The company's shareholders are another critical stakeholder group because they ultimately give or withhold their approval of management's decisions through their proxies. Moreover, their "buy" and "sell" decisions influence the corporation's stock price.

Government regulatory agencies, another important set of stakeholders, are required by law to monitor and regulate firms for specific, publicly defined purposes. Nevertheless, these agencies have considerable discretion in how they interpret and apply the law. Where they have a positive perception of the firm, they are likely to be much less censorious.

Social action organizations represent still another set of stakeholders. To the extent a corporation has a negative reputation in the particular area of concern of a social action group; it likely will be targeted for criticism and harassment by that group. For example, the Labor/Community Strategy Center has organized aboycott of Texaco stations and products in an effort to influence the company to reduce the air pollution emanating from its refinery in Wilmington, California. Although there are many refineries in the Wilmington area, the environmental group targeted Texaco for its boycott because of a recent much publicized explosion at the company's refinery.

A strong positive image with the general public can be beneficial to the firm. Research suggests that a prominent corporate image and an outstanding reputation are consequential factors in attracting a high qualityworkforce. Merck, Microsoft, and Hewlett-Packard, for instance, have traditionally attracted topnotch job applicants because of their sterling reputations.

Current employees represent the internal constituency that a firm must consider when communicating corporate identity. It is widely believed that a positive reputation in the eyes of employees is a prime causal factor of high morale and productivity. This condition is frequently cited as a fundamental reason for the success of Japanese firms. Additionally, it should be emphasized that employees play a large role in representing the company to its external stakeholders.

Obviously, each of the various stakeholder groups is likely to have a somewhat different perception of the corporation because each is concerned primarily with a different facet of its operation. Thus, customers are principally interested in the price, quality, and reliability of the company's products and services. Financial institutions are concerned with financial structure and performance. Employees are mainly concerned with wages, working conditions, and personnel policies. Logically, then, a company should tailor its communication to each stakeholder group individually to engage the special concerns of that group.

A consistent image among the various stakeholder groups, however, is also essential. Although it is prudent to stress different facets of the firm's identity to its various publics, the firm should avoid projecting an inconsistent image for two key reasons. First, some of the concerns of the stakeholders overlap. For example, the financial community and the shareholders would have many of the same financial and strategic concerns about the company. In fact, many shareholders rely heavily on the advice of experts from financial institutions. Both employees and the general public have an interest in the overall prestige of the firm and the reputation of its products. A social action group's criticism, as in the case of the Texaco boycotts, whether economically effective or not, is bound to influence some customers and affect the company's public reputation. Of course, a regulatory agency such as the Occupational Safety and Health Administration would focus narrowly on the firm's safety record and policies but the company's employees and their labor unions also have a stake in these matters.

The second and related reason for avoiding an inconsistent image is that the sundry stakeholders are not separate, discrete entities. Membership overlaps. Consider the example of a typical public utility where almost all of its employees are also customers and a significant number may also be shareholders. Furthermore, it is not unlikely that some of its employees will be active in environmental or consumer rights groups that challenge the company on specific issues. It is also likely that some of the company's bankers and regulators will be among its customers.

Corporate Communication

Corporate communication is the link between corporate identity and corporate image and reputation. It should be defined in the broadest possible sense because companies communicate their identities in many different ways. This includes almost everything they do from the way telephones are answered to the involvement of their employees in community affairs. A description of the principal communication media is presented below.

Nomenclature

The primary concerns in this category are the names used to identify the corporation, its divisions, and its products. In recent years, many firms have changed their corporate names to communicate a major change in identity. To illustrate, International Harvester changed its name to Navistar to signal its exit from the agricultural equipment industry. Carter Hawley Hale Stores changed its name to Broadway Stores to identify more closely with its store operations and accentuate its revitalization since emerging from bankruptcy.

Graphics

Graphics, which were the original focus of image consultants, are concerned with the overall visual presentation of the organization. The graphics system should dictate the design style of the company's literature, signs, and stationery. It involves coordinating the style of the typeface, photography, illustrations, layout, and coloring in all the company's graphics. The key question here, as with the nomenclature issue, is whether the company's visual presentation is appropriately communicating its identity. Consider the example of Alitalia Airline. Although Alitalia was one of the largest transatlantic carriers, it projected an image of a relatively small, casual, inefficient "Italian" airline. To counter this negative image, Alitalia, following the lead of Olivetti and Ferrari, developed a graphics program stressing superior design and high technology. All forms of corporate communication such as aircraft insignias, uniforms, baggage tags, and promotional materials were redesigned to consistently project and reinforce this positive image. Today Alitalia is regarded by the flying public as a major global carrier.

The logo is the heart of the corporate graphics design system. Unlike nomenclature, logos can be changed subtly over time to reflect the evolving corporate identity. For example, Shell has varied its graphics system many times over the past century. Through all its changes, the company retained, for the sake of continuity, some version of its basic seashell logo. In contrast, Transamerica Corporation abruptly replaced its recognizable "T" logo with the "Transamerica Pyramid" (a well-known San Francisco landmark) to signal its metamorphosis from a conglomerate to a focused financial services company.

Formal Statements

This category includes mission statements, credos, and codes of ethics, annual reports, advertising copy, and company slogans. Company slogans can be a particularly potent means of communicating to stakeholders. Avis's "We Try Harder", for instance, has been remarkably effective in conveying the company's identity. Other examples include Prudential's "Own a Piece of the Rock", which communicates the company's financial stability, and E. I. du Pont de Nemours & Co.'s "Better Living through Chemistry", which underscores the firm's science-driven culture.

Architecture

The design of corporate buildings and the interior layout of offices and factories also can reveal much about a company. A series of closed offices suggests a very different culture from a large open room with desks in full sight of each other. The Transamerica Pyramid as a symbol of financial services illustrates the potential for communicating through architecture. SmithKline Beecham (the resultant company from the merger of the pharmaceutical firms Beecham and SmithKline Beckman) specifically selected a corporate headquarters complex in London that projects a culture that it hopes will evolve at the merged company.

Interactions and Events

This is a catch-all category, but a critical one, because every interaction a company employee has with a stakeholder, and every event related to a company, communicates something of the firm's identity. This means, for one thing, that employees should be trained and motivated to project a positive image of the company. The increased popularity of training employees on answering telephones shows that many firms understand the criticality of this communication source.

Unexpected events also can conspicuously communicate corporate identity. As mentioned above, the *Exxon Valdez* oil spill became a dimension of the Exxon Corporation's identity, but it also immediately imparted the image of an environmentally insensitive firm to most, if not all, of the company's stakeholders. In similar fashion, the catastrophe at Union Carbide's Bhopol plant projected a negative image, as did the controversy over the treatment of African-American customers at Denny's. A company's reaction to such events, however, also can play a prominent role in its projected image. The failure of Chairman Rawls to travel to the site of the oil spill further compounded Exxon's image problems. Conversely, Johnson & Johnson's prompt nationwide recall of Tylenol bottles after Chicago-area deaths were attributed to a few poisoned Tylenol capsules significantly reduced the negative impact of these tragedies on the company. In fact, many observers believe that Johnson & Johnson's image and its reputation for social responsiveness has emerged stronger than ever.

Feedback

Feedback is essential to managing the corporate image. Without it, company executives are "flying blind". They need accurate information on stakeholder perceptions if they are to make sound decisions. Ideally, feedback should be continuous. As a practical matter, relatively continuous feedback can be elicited from salespeople, public relationsexecutives, finance managers, and other employees who routinely interact with stakeholders. Based on such input, modifications may be made in the company's communication methods or, if warranted, a formal study of the corporate identity initiated. In addition to systematically utilizing internal sources, it is prudent to conduct formal studies on a regular basis, say every five years. Formal studies are typically performed by identity/image consultants using in-depth, one-on-one and group interviews as their chief research tools. This type of comprehensive outside review would normally include an analysis of the corporate identity, an appraisal of the firm's image and reputation in the eyes of its stakeholders, and an evaluation of the efficacy of its corporate communications. The consultant's recommendations might run from making slight alterations in the corporate communication program to a reshaping of the firm's identity.

Two examples illustrate the importance of feedback and taking appropriate remedial measures based on the feedback. In the first example, Consolidated Foods found that it had a deficient and inaccurate image in the financial community. Specifically, it learned that its rather bland sounding name translated into a bland image in the eyes of financial analysts. The company also discovered that not only was its name bland, but it also was inappropriate because its organization was decentralized and encompassed a variety of business units; therefore, it could not be accurately described as "consolidated". To remedy the situation, the company renamed itself Sara Lee after its most prestigious product line. Subsequent research showed that awareness of the company among financial analysts increased significantly.

In the second example, Jaguar, in the days prior toprivatization, learned from research that it had a terrible reputation for quality and reliability among customers. To correct this problem, Jaguar initiated a rigorous quality program which has helped the firm regain its earlier reputation for quality vehicles.

Conclusions

In the past, corporate identity was seen almost universally as the narrow, peripheral function of graphic design. Today, however, as a consequence of the epochal, often abrupt changes occurring throughout the business world, and the resultant danger of corporate images becoming outmoded and erroneous and reputations deteriorating, the issue of managing the corporate identity has been elevated to a level of strategic importance in executive circles.

The modern concept of corporate identity has a broad sweep and a strategic focus. It views a company's image and reputation among its several stakeholders as critical resources over which the firm has control. The framework presented here outlines a conceptual model through which management can comprehend, monitor, and influence the development of these intangible assets. The concept is relatively simple but its effective implementation can be profoundly challenging. The firms that master this challenge will, in all likelihood, be the ones that will survive and prosper today's turbulent business environment.

Unit Three
The Third Wave and Post-Modernism

Core Text 8

Development of the Information Society Model

There is currently no universally accepted concept of what exactly can be termed information society and what shall rather not so be termed. Most theoreticians agree that a transformation can be seen that started somewhere between the 1970s and today and is changing the way societies work fundamentally. Information technology is not only Internet, and there are discussions about how big the influence of specific media or specific modes of production really is.

Some people, such as Antonio Negri and Newt Gingrich, characterize the information society as one in which people do immaterial labour. By this, they appear to refer to the production of knowledge or cultural artifacts. One problem with this model is that it ignores the material and essentially industrial basis of the society. However it does point to a problem for workers, namely how many creative people does this society need to function? For example, it may be that you only need a few star performers, rather than a plethora of non-celebrities, as the work of those performers can be easily distributed, forcing all secondary players to the bottom of the market. It is now common for publishers to promote only their best selling authors and to try to avoid the rest—even if they still sell steadily. Films are becoming more and more judged, in terms of distribution, by their first weekend's performance, in many cases cutting out opportunity for word-of-mouth development.

Another problem with the idea of the information society is that there is no easily agreed upon definition of the term, which can not only include art, texts, blueprints and scientific theories, but also lies, football results, trivia, random letters, mistakes and so on. Information is not necessarily productive or useful. It can even be harmful.

Considering that metaphors and technologies of information move forward in a reciprocal relationship, we can describe some societies (especially the Japanese society) as an information society because we think of it as such.

Second and Third Nature

As mentioned earlier an information society is the means of getting information from one place to another. As technology has become more advanced over time so too has the way we have adapted in sharing this information with each other.

"Second nature" refers to a group of experiences that get made over by culture. They then get remade into something else that can then take on a new meaning. As a society we transform this process so it becomes something natural to us, i.e second nature. So, by following a particular pattern created by culture we are able to recognize how we use and move information in different ways. From sharing information via different time zones (such as talking online) to information ending up in a different location (sending a letter overseas) this has all become a habitual process that we as a society take for granted.

However, through the process of sharing information vectors have enabled us to spread information even further. Through the use of these vectors information is able to move and then separate from the initial things that enabled them to move. From here, something called "third nature" has developed. An extension of second nature, third nature is in control of second nature. It expands on what second nature is limited by. It has the ability to mould information in new and different ways. So, third nature is able to "speed up, proliferate, divide, mutate, and beam in on us from else where". It aims to create a balance between the boundaries of space and time (see second nature). This can be seen through the telegraph. It was the first successful technology that could send and receive information faster than a human being could move an object. As a result different vectors of people have the ability to not only shape culture but create new possibilities that will ultimately shape society.

Therefore, through the use of second nature and third nature society is able to use and explore new vectors of possibility where information can be moulded to create new forms of interaction.

Related Terms

A number of terms in current use emphasize related but different aspects of the emerging global economic order. The Information Society intends to be the most encompassing in that an economy is a subset of a society. The Information Age is somewhat limiting, in that it refers to a 30-year period between the widespread use of computers and the knowledge economy, rather than an emerging economic order. The knowledge era is about the nature of the content, not the socioeconomic processes by which it will be traded. The computer revolution and knowledge revolution refer to specific revolutionary transitions, rather than the end state towards which we are evolving. The Information Revolution relates with the well known terms agricultural revolution and industrial revolution.

The information economy and the knowledge economy emphasize the content or intellectual property that is being traded through an information market or knowledge market, respectively. Electronic commerce and electronic business emphasize the nature of transactions and running a business, respectively, using the Internet and World-Wide Web. The digital economy focuses on trading bits in cyberspace rather than atoms in physical space. The network economy stresses that businesses will work collectively in webs or as part of business ecosystems rather than as stand-alone units. Social networking refers to the process of collaboration on massive, global scales. The Internet economy focuses on the nature of markets that are enabled by the Internet. Knowledge services and knowledge value put content into an economic context. Knowledge services integrate knowledge management, within a knowledge organization, that trades in a knowledge market.

Although seemingly synonymous, each term conveys more than nuances or slightly different views of the same thing. Each term represents one attribute of the likely nature of economic activity in the emerging post-industrial society. Alternatively, the new economic order will incorporate all of the above plus other attributes that have not yet fully emerged.

Intellectual Property Considerations

One of the central paradoxes of the information society is that it makes information easily reproducible, leading to a variety of freedom/control problems relating to intellectual property. Essentially, business and capital, whose place becomes that of producing and selling information and knowledge, seems to require control over this new resource so that it can effectively be managed and sold as the basis of the information economy. However, such control can prove to be both technically and socially problematic. Technically because copy protection is often easily circumvented and socially rejected because the users and citizens of the information society can prove to be unwilling to accept such absolute commodification of the facts and information that compose their environment.

Responses to this concern range from the Digital Millennium Copyright Act in the United States (and similar legislation elsewhere) which make copy protection circumvention illegal, to the free software, open source and copyleft movements, which seek to encourage and disseminate the "freedom" of various information products (traditionally both as in "gratis" or free of cost, and liberty, as in freedom to use, explore and share).

Caveat: Information society is often used by politicians meaning something like "we all do Internet now"; the sociological term information society (or informational society) has some deeper implications about change of societal structure.

Key Words

[1] information society 信息社会

[2] immaterial [ɪməˈtɪərɪəl] adj. 1. 不重要的，不相干的 2. 非物质的，无形的

[3] word-of-mouth adj. 口头的，口述的

[4] metaphor [ˈmetəfə] n. 隐喻

[5] information revolution 信息革命

[6] synonymous [sɪˈnɒnɪməs] adj. 同义的，类义的

[7] intellectual property 知识产权

Key Sentences

1. Another problem with the idea of the information society is that there is no easily agreed upon definition of the term, which can not only include art, texts, blueprints and scientific theories, but also lies, football results, trivia, random letters, mistakes and so on. Information is not necessarily productive or useful. It can even be harmful.

有关信息社会思想的另一个难题是有关这一术语的定义不容易达成共识。这个概念不只包括艺术、文本、设计图和科学理论，还有谎言、足球赛结果、不足道的细枝末节、随意的字母、错误等。信息不是必须多产的或有用的，它甚至是有害的。

2. The information economy and the knowledge economy emphasize the content or intellectual property that is being traded through an information market or knowledge market, respectively.

信息经济和知识经济强调通过信息市场或知识市场各自进行内容交易或知识产权交易。

课文翻译

信息社会模式的发展

目前，信息社会对于什么是信息社会，什么又不是信息社会还没有一个普遍认同的确切定义。大部分理论家普遍认为，从20世纪70年代的某个时期开始至如今有一个显而易见的变化，并且社会工作方式正在从根本上发生转变。信息技术不仅是互联网，还有关于特定媒体或特定生产模式的影响有多大的讨论。

有些人，如安东尼·内格里和纽特·金里奇，描述信息社会具有人们进行非物质劳动的特征。通过这一点，他们似乎指的是知识或文化产品生产。这个观点的一个问题是它忽略了社会物质和必需的工业基础。但它确实为劳动者指出一个问题，即在这个社会需要有多少创造性的人们发挥作用？例如，可能你只需要一些明星的表演，而不是过多的非名人，因为这些表演者的工作能很容易被分散，迫使所有二级演员处于市场底层。只提拔销量最好的作者并且远离其他作者，即使他们的作品还在稳步销售，这对出版商来说是很普遍的。电影正越来越多地根据发行量被评价，通过其第一个周末的表现，在许多情况下会切断口碑营销发展的机会。

信息社会思想的另一个难题是有关这一术语的定义不容易达成共识，这个概念不只包括艺术、文本、设计图和科学理论，还有谎言、足球赛结果、不足道的细枝末节、随意的字母、错误等。信息不是必须多产的或有用的，它甚至是有害的。

考虑到隐喻和信息技术以一种互惠关系向前发展，我们可以描述一些社会（特别是日本社会）是信息社会，因为我们认为它是这样。

第二和第三特性

正如前面所提到的，信息社会意味着从一个地方到另一个地方获取信息。随着时间的推移，技术日益先进，我们已经习惯了彼此分享信息的方式。

"第二特性"是指通过文化改进得到的一组经验，它们接下来重新形成一些具有新含义的观念。作为一个社会，我们改变这个形成的过程，使其对我们来说变得自然，即第二特性。因此，通过跟随由文化创造的特定模式，我们能够认识到如何以不同的方式使用和传递信息，通过不同的时区共享信息（如在线聊天）到在不同的位置结束信息（发送海外信件）都成为一个习惯的过程，我们认为这个社会理所当然是这样。

然而，分享信息载体的过程使我们能够进一步传播信息。通过这些载体的使用，信息能够传递，并与能使它们传递的初期事物区分出来。至此，被称为"第三特性"的事物已经形成了。作为第二特性的一个延伸，第三特性控制着第二特性。它补充了第二特性所限制的部分。它有能力以新的和不同的方式塑造信息。因此，第三特性是能够"加快、增殖、分化、变异以及从其他地方影响我们。"它的目标是在空间和时间界限之间建立平衡（见第二特性），以电报为例，它是第一个比人类传递物体更快的，能够发送和接收信息的成功技术。因此，人们的不同载体不仅能够塑造文化形态，而且能够产生新的最终塑造社会形态的可能性。

因此，通过第二特性和第三特性的使用，社会能够使用和开发可能的新载体，通过这些载体信息被塑造而产生新形式的交互。

相关术语

目前使用的许多术语被用来强调相关但又不同的新兴全球经济秩序的各种现象。信息社会包罗万象，经济只是其中的一个子集。信息时代具有一定的局限性，因为它指的是由电脑广泛使用至知识经济时代之间的一个30年时期，而不是一个新兴经济秩序。知识经济时代则包含了这方面的内容，它将不按照常规的社会经济程序进行交易。计算机变迁和知识变迁指的是一个特殊变迁的过渡时期，而不是我们为此不断发展的终极状态。信息革命可以同农业革命和工业革命等众所周知的术语联系起来。

信息经济和知识经济强调通过信息市场或知识市场各自进行内容交易或知识产权交易。电子商务和电子商业强调分别使用因特网和万维网处理事务和运行业务的本质。数字经济的焦点是交易在比特单位的网络空间进行，而不是在原子单位的物理空间进行。网络经济强调企业群体将共同在网络或作为商业生态系统的一部分进行运营，而不是作为独立经营单元进行运营。社会化网络指大规模的、在全球范围内的协作过程。互联网经济的焦点是市场交易本质上是通过互联网进行的。知识服务和知识价值将内容集中于经济范畴。知识服务整合了知识组织内进行知识市场交易的知识管理。

虽然看起来意义相似，每个术语包含了更多的细微差别或对同一事物的轻微不同看法。每个术语代表在后工业社会中经济活动的一种本质属性。另外，新的经济秩序将结合上述以及其他尚未完全出现的其他特征。

知识产权的思考

信息社会的一个首要悖论是，它使信息容易复制，从而导致各种关于知识产权自主/控制的问题。从本质上讲，企业和中心城市，这些成为生产和销售信息和知识的地方，似乎都需要对这个新的资源进行控制，以便作为信息经济的基础能进行有效的管理和销售。但是，这种控制可能会带来技术和社会两方面的问题。从技术上讲，因为版权保护往往容易被规避和被社会拒绝，因为信息社会的用户和社会公民不愿意接受这种构成其环境的信息和商品化的事实。

对于这些关注的反应，从使规避复制保护非法的美国千禧年数字版权法（与其他相似立法）到自由软件、开放资源与非盈利版权的运动，它将致力于鼓励和宣传各类信息产品的"自由"（传统上叫"免费"或自由成本和许可权，也叫自由使用、开发和共享）。

附加说明：信息社会经常被政客们用来意指如"现在大家都在互联网上"等事情，社会学的术语"信息社会（或信息化社会）"则进一步暗示了有关社会结构的变化。

Free Reading 1

Tools of Tomorrow

Coal, rail, textile, steel, auto, rubber, machine tool manufacture—these were the classical industries of the Second Wave. Based on essentially simple electromechanical principles, they used high energy inputs, spat out enormous waste and pollution, and were characterized by long production runs, low skill requirements, repetitive work, standardized goods, and heavily centralized controls.

From the mid-1950's it became increasingly apparent that these industries were backward and waning in the industrial nations. In the United States, for example, while the labor force grew by 21 percent between 1965 and 1974, textile employment rose by only 6 percent and employment in iron and steel actually dropped 10 percent a similar pattern was evident in Sweden, Czechoslovakia, Japan, and other Second Wave nations.

As these old-fashioned industries began to be transferred to developing countries, where labor was cheaper and technology less advanced, their social influence also began to die out and a set of dynamic new industries shot up to take their place.

These new industries differed markedly from their predecessors in several respects: they were no longer primarily electromechanichal and no longer based on the classic science of the Second Wave era. Instead, they rose from accelerating breakthroughs in a mix of scientific disciplines that were rudimentary or even nonexistent as recently as twenty-five years ago quantum electronics, information theory, molecular biology, oceanics, nucleonics, ecology, and the space sciences. And they made it possible for us to reach beyond the grosser features of time and space with which Second Wave industry concerned itself, to manipulate, as Soviet physicist B. G. Kuznetsov has noted, "very small spatial regions (say, of the radius of an atomic nucleus, i. e. , 1(H3centimeters) and temporal intervals of the order of Kh23 seconds. "

It is from, these new sciences and our radically enhanced manipulative abilities, that the new industries arose—computers and data processing, aerospace, sophisticated petrochemicals, semiconductors, advanced communications, and scores of others.

In the United States, where this shift from Second Wave to Third Wave technologies began earliest—sometime in the mid-1950's—old regions like the Merrimack Valley in New England sank into the status of depressed areas while places like Route 128 outside Boston or "Silicon Valley" in California zoomed into prominence, their suburban homes filled with specialists in solid-state physics, systems engineering, artificial intelligence, or polymer chemistry.

Moreover, one could track the transfer of jobs and affluence as they followed the transfer of technology, so that the so-called "sun-belt" states, fed by heavy defense contracts, built an advanced technological base while the older industrial regions in the Northeast and around the Great Lakes plunged into lassitude and near-bankruptcy. The long running financial crisis of New York City was a clear reflection of this technological upheaval. So, too, was the stagnation of Lorraine, France's center of steelmaking. And so, at yet another level was the failure of British socialism. Thus, at the end of World War II the Labour government spoke of seizing the "commanding heights" of industry and did so. But the commanding heights it nationalized turned out to be coal, rail, and steel—precisely those industries being by-passed by the technological revolution: yesterday's commanding heights.

Today, cheap mini-computers are about to invade the American home. By June 1979 some one hundred companies were already manufacturing home computers. Giants like Texas Instruments were in the field, and chains like Sears and Montgomery Ward were on the edge of adding computers to their household wares. "Some day soon," chirruped a Dallas microcomputer retailer, "every home will have a computer. It will be as standard as a toilet."

Linked to banks, stores, government offices, to neighbors' homes and to the workplace, such computers are destined to reshape not only business, from production to retailing, but the very nature of work and, indeed, even the structure of the family.

Like the computer industry to which it is umbilically tied, the electronics industry has also been exploding, and consumers have been deluged with hand-held calculators, diode watches, and TV-screen games. These, however, provide only the palest hint of what lies in store: tiny, cheap climate and soil sensors in agriculture; infinitesimal medical devices built into ordinary clothing to monitor heartbeat or stress levels of the wearer—these and a multitude of other applications of electronics lurk just beyond the present.

The advance toward Third Wave industries, moreover, will be radically accelerated by the energy crisis, inasmuch as many of them carry us toward processes and products that are miserly in their energy requirements. Second Wave telephone systems, for example, required virtual copper mines beneath the city streets—endless miles of snaking cable, conduit, relays, and switches. We are now about to convert to fiber optic systems that uses hair-thin light-carrying fibers to convey messages. The energy implications of this switchover are staggering: it takes about one thousand the energy to manufacture optical fiber that it took to dig, smelt, and process an equivalent length of copper wire. The same ton of coal required to produce 90 miles of copper wire can turn out 80,000 miles of fiber!

The shift to solid-state physics in electronics moves in the same direction, each step forward producing components that require smaller and smaller inputs of energy, At IBM, the latest developments in L.S.I. (Large Scale Integration) technology involve components that are activated by as little as fifty microwatts.

This characteristic of the electronic revolution suggests that one of the most powerful conservation strategies for energy-starved high technology economies may well be the rapid substitution of low energy Third Wave industries for energy wasting Second Wave industries.

More generally, the journal Science is correct when it states that "the country's economic activity may be substantially altered" by the electronics explosion. "Indeed, it is probable that reality will outstrip fiction in the rate of introduction of new and often unexpected applications of electronics."

The electronics explosion, however, is only one step in the direction of an entirely new techno-sphere.

Free Reading 2

The Electronic Cottage

Doing Homework

Yet there were equally, if not more, compelling reasons three hundred years ago to believe people would never move out of the home and field to work in factories. After all, they had labored in their own cottages and the nearby land for 10,000 years, not a mere 300. The entire structure of family life, the process of child-rearing and personality formation, the whole system of property and power, the culture, the daily struggle for existence were all bound to the hearth and the soil by a thousand invisible chains. Yet these chains were slashed in short order as soon as a new system of production appeared.

Today that is happening again, and a whole group of social and economic forces are converging to transfer the locus of work.

To begin with, the shift from Second Wave manufacturing to the new, more advanced Third Wave manufacturing reduces, as we just saw, the number of workers who actually have to manipulate physical goods. This means that even in the manufacturing sector an increasing amount of work is being done that—given the right configuration of telecommunication and other equipment—could be accomplished anywhere, including one's own living room. Nor is this just a science fiction fantasy.

When Western Electric shifted from producing electromechanical switching equipment for the phone company to making electronic switching gear, the work force at its advanced manufacturing facility in northern Illinois was transformed. Before the changeover, production workers outnumbered white-collar and technical workers three to one. Today the ratio is one to one. This means that fully half of the 2,000 workers now handle information instead of things, and much of their work can be done at home. Dom Cuomo, director of engineering at the Northern Illinois facility, put it flatly, "If you include engineers, ten to twenty-five percent of what is done here could be done at home with existing technology."

Indeed, an unmeasured but appreciable amount of work is already being done at home by such people as salesmen and saleswomen who work by phone or visit, and only occasionally touch base at the office; by architects and designers; by a burgeoning pool of specialized consultants in many industries; by large numbers of human-service workers like therapists or psychologists; by music teachers and language instructors; by art dealers, investment counselors, insurance agents, lawyers, and academic researchers; and by many other categories of white-collar, technical, and professional people.

These are, moreover, among the most rapidly expanding work classifications, and when we suddenly make available technologies that can place a low-cost "work station" in any home, providing it with a "smart" typewriter, perhaps, along with a facsimile machine or computer console and teleconferencing equipment, the possibilities for home work are radically extended.

In short, as the Third Wave sweeps across society, we find more and more companies that can be described, in the words of one researcher, as nothing but "people huddled around a computer." Put the computer in people's homes, and they no longer need to huddle. Third Wave white-collar work, like Third Wave manufacturing, will not require 100 percent of the work force to be concentrated in the workshop.

One should not underestimate the difficulties entailed in transferring work from its Second Wave locations in factory and office to its Third Wave location in the home. Problems of motivation and management, of corporate and social reorganization will make the shift both prolonged and, perhaps, painful. Nor can all communication be handled vicariously. Some jobs—especially those involving creative deal-making, where each decision is nonroutine require much face-to-face contact. Thus Michael Koerner, President of Canada Overseas Investments, Ltd., says, "We all need to be within a thousand feet of one another."

The Telecommuters

Nevertheless, powerful forces are converging to promote the electronic cottage. The most immediately apparent is the economic trade-off between transportation and telecommunication. Most high-technology nations are now experiencing a transportation crisis, with mass transit systems strained to the breaking point, roads and highways clogged, parking spaces rare, pollution a serious problem, strikes and breakdowns almost routine, and costs skyrocketing.

But these are not the only forces subtly moving us toward the geographical dispersal of production and, ultimately, the electronic cottage of the future. The Nilles team found that the average American urban commuter uses the gasoline equivalent of 64.6 kilowatts of energy to get back and forth to work each day. (The Los Angeles insurance employees burned 37.4 million kilowatts a year in commuting.) By contrast, it takes far less energy to move information.

A typical computer terminal uses only 100 to 125 watts or less when it is in operation, and a phone line consumes only one watt or less while it is in use. Making certain assumptions about how much communications equipment would be needed, and how long it would operate, Nilles calculated that "the relative energy consumption advantage of telecommuting over commuting (i.e., the ratio of commuting energy consumption to telecommuting consumption) is at least 29∶1 when the private automobile is used; 11∶1 when normally loaded mass transit is used; and 2∶1 for 100 percent utilized mass transit systems."

To this we can add even more pressures tending in the same direction. Corporate and government employers will discover that shifting work into the home—or into local or neighborhood work centers as a halfway measure—can sharply reduce the huge amounts now spent for real estate. The smaller the central offices and manufacturing facilities become, the smaller the real estate bill, and the smaller the costs of heating, cooling, lighting, policing, and maintaining them. As land, commercial and industrial real estate, and the associated tax load all soar, the hope of reducing and/or externalizing these costs will favor the farming-out of work.

The transfer of work and the reduction of commuting will also reduce pollution and therefore the tab for cleaning it up. The more successful environmentalists become at compelling companies to pay for their own pollution, the more incentive there will be to shift to low-polluting activities, and therefore from large-scale, centralized workplaces to smaller work centers or, better yet, into the home.

Social factors, too, support the move to the electronic cottage. The shorter the workday becomes, the longer the commuting time in relationship to it. The employee who hates to spend an hour getting to and from the job in order to spend eight hours working may very well refuse to invest the same commuting time if the hours spent on the job are cut. The higher the ratio of commuting time to working time, the more irrational, frustrating, and absurd the process of shuttling back and forth. As resistance to commuting rises, employers will indirectly have to increase the premium paid to workers in the big, centralized work locations, as against those willing to take less pay for less travel time, inconvenience, and cost. Once again there will be greater incentive to transfer work.

Finally, deep value changes are moving in the same direction. Quite apart from the growth of privatism and the new allure of small-city and rural life, we are witnessing a basic shift in attitude toward the family unit. The nuclear family, the standard, socially approved family form throughout the Second Wave period, is clearly in crisis. For now, we need only note that in the United States and Europe—wherever the transition out of the nuclear family is most advanced—there is a swelling demand for action to glue the family unit together again. And it is worth observing that one of the things that has bound families tightly together through history has been shared work.

Even today one suspects that divorce rates are lower among couples who work together. The electronic cottage raise once more on a mass scale the possibility of husbands and wives, and perhaps even children, working together as a unit. And when campaigners for family life discover the possibilities inherent in the transfer of work to the home we may well see a rising demand for political measures to speed up the process—tax incentives, for example, and new conceptions of workers rights.

During the early days of the Second Wave era, the workers' movement fought for a "Ten Hour Day", a demand that would have been almost incomprehensible during the First Wave period. Soon we may see the rise of movements demanding that all work that can be done at home be done at home. Many workers will insist on that option as a right. And, to the degree that this relocation of work is seen as strengthening family life, their demand will receive strong support from people of many different political, religious, and cultural persuasions.

The fight for the electronic cottage is part of the larger super-struggle between the Second Wave past and the Third Wave future, and it is likely to bring together not merely technologists and corporations eager to exploit the new technical possibilities but a wide range of other forces—environmentalists, labor reformers of a new style, and a broad coalition of organizations, from conservative churches to radical feminists and mainstream political groups—in support of what may well be seen as a new, more satisfactory future for the family. The electronic cottage may thus emerge as a key rallying point of the Third

Wave forces of tomorrow.

The Home-Centered Society

If the electronic cottage were to spread, a chain of consequences of great importance would flow through society. Many of these consequences would please the most ardent environmentalist or techno-rebel, while at the same time opening new options for business entrepreneurship.

Community Impact: Work at home involving any sizeable fraction of the population could mean greater community stability—a goal that now seems beyond our reach in many high-change regions. If employees can perform some or all of their work tasks at home, they do not have to move every time they change jobs, as many are compelled to do today. They can simply plug into a different computer.

This implies less forced mobility, less stress on the individual, fewer transient human relationships, and greater participation in community life. Today when a family moves into a community, suspecting that it will be moving out again in a year or two, its members are markedly reluctant to join neighborhood organizations, to make deep friendships, to engage in local politics, and to commit themselves to community life generally. The electronic cottage could help restore a sense of community belonging, and touch off a renaissance among voluntary organizations like churches, women's groups, lodges, clubs, athletic and youth organizations. The electronic cottage could mean more of what sociologists, with their love of German jargon, call gemeinschaft.

Environmental Impact: The transfer of work, or any part of it, into the home could not only reduce energy requirements, as suggested above, but could also lead to energy decentralization. Instead of requiring highly concentrated amounts of energy in a few high-rise offices or sprawling factory complexes, and therefore requiring highly centralized energy generation, the electronic cottage system would spread out energy demand and thus make it easier to use solar, wind, and other alternative energy technologies. Small-scale energy generation units in each home could substitute for at least some of the centralized energy now required. This implies a decline in pollution as well, for two reasons: first, the switch to renew-able energy sources on a small-scale basis eliminates the need for high-polluting fuels, and second, it means smaller releases of highly concentrated pollutants that overload the environment at a few critical locations.

Economic Impact: Some businesses would shrink in such a system, and others proliferate or grow. Clearly, the electronics and computer and communications industries would flourish. By contrast, the oil companies, the auto industry, and commercial real estate developers would be hurt. A whole new group of small-scale computer stores and information services would spring up; the postal service, by contrast, would shrink. Papermakers would do less well; most service industries and white-collar industries would benefit.

At a deeper level, if individuals came to own their own electronic terminals and equipment, purchased perhaps on credit, they would become, in effect, independent entrepreneurs rather than classical employees—meaning, as it were, increased ownership of the "means of production" by the worker. We might also see groups of home workers organize themselves into small companies to contract for their services or, for that matter, unite in cooperatives that jointly own the machines. All sorts of new relationships and organizational forms become possible.

Psychological Impact: The picture of a work world that is increasingly dependent upon abstract symbols conjures up an overcerebral work environment that is alien to us and, at one level, more impersonal than at present. But at a different level, work at home suggests a deepening of face-to-face and emotional relationships in both the home and the neighborhood. Rather than a world of purely vicarious human relationships, with an electric screen interposed between the individual and the rest of humanity, as imagined in many science fiction stories, one can postulate a world divided into two sets of human relationships—one real, the other vicarious—with different rules and roles in each.

It is not possible to see in relationship to one another a number of Third Wave changes usually examined in isolation. We see a transformation of our technological system and our energy base into a new techno-sphere. This is occurring at the same time that we are de-massifying the mass media and building an intelligent environment, thus revolutionizing the info-sphere as well. In turn, these two giant currents flow together to change the deep structure of our production system, altering the nature of work in factory and office and, ultimately, carrying us toward the transfer of work back into the home.

By themselves, such massive historical shifts would easily justify the claim that we are on the edge of a new civilization. But we are simultaneously restructuring our social life as well, from our family ties and friendships to our schools and corporations. We are about to create, alongside the Third Wave techno-sphere and info-sphere, a Third Wave socio-sphere as well.

Core Text 9

Pop Design

The term Pop was coined in the 1950s and referred to the emergence of popular culture during that decade. In 1952, the Independent Group was founded in London and its members, including the artist Richard Hamilton (b. 1922). The sculptor Eduardo Paolozzi (b. 1924), the design critic Reyner Banham (1920-1988) and the architects, Peter and Alison Smithson, were among the first to explore and celebrate the growth of popular consumer culture in America.

In the 1960s, American artist too, such as Andy Warhol(1928—1987), Roy Lichtenstein (1923—1998) and Claes Oldenburg (b. 1929) began drawing inspiration from the "low art" aspects of contemporary life such as advertising, packaging, comics and television.

Not surprisingly, pop also began to manifest itself in the design of everyday use, as designer sought a more youth-based and less serious approach than had been offered by the Good Design of the 1950s. The ascendancy of product styling in the 1950s, in the name of productivity-increasing built-in obsolescence, provided fertile ground for the "use-it-today, sling-it-tomorrow" ethos that permeated industrial production during the 1960s.

Figure 3.1 Peter-Murdoch, Spotty, 1963 Polyethylene-coated, laminated paperboard construction Murdoch's polka-dot child's chair "Spotty" is an icon. Of the Pop era, its low production costs and inherent disposability were ideally suited to the demands of the mass consumer market.

Peter Murdoch's polka dotted cardboard Spotty child's chair (shown in Figure 3.1) and De Pas, D'urbino and Lomazzi's PVC Blow chair (1967) were eminently disposable and epitomized the widespread culture of ephemerality. So too did the plethora of short-lived gimmicks such as paper dresses, which were lauded for their novelty in the large number of color supplements and glossy magazines that became increasingly dependent on featuring such items.

For many designers working within the Pop idiom, plastics became their materials of choice. By the 1960s, many new types of plastics and aligned processes, such as injection-moulding, became available and relatively inexpensive to use. The bright rainbow colors and bold forms associated with Pop Design swept the last vestiges of post-war austerity and reflected the widespread optimism of the 1960s, which was bolstered by unprecedented economic prosperity and sexual liberation. Since Pop Design was aimed at the youth-market, products had to be cheap and were therefore often of poor quality. The expendability of such products, however, became part of their appeal as they represented the antithesis of the "timeless" modern classics that had been promoted in the 1950s.

Pop Design with its Anti-Design association countered the Modern Movement's sober dictum "Less is More" and led directed to the Radical Design of the 1970s. It drew inspiration from a wide range of sources—Art Nouveau, Art Deco, Futurism, Surrealism, Op Art, Psychedelia, Eastern Mysticism, Kitsch and the Space-Age—and was spurred on by the growth of the global mass-media.

The oil crisis of the early 1970s, however, necessitated a more rational approach to design and Pop Design was replaced by the Craft Revival on the one hand and High Tech on the other. By questioning the precept of Good Design, and thereby Modernism, the influence of Pop Design was far-reaching and laid some of the foundations on which post-modernism was to grow. (Some representative works are shown in Figure 3.2—Figure 3.5)

Figure 3.2　Gaetano Pesce, Up Series for C&B Italia, 1969 Gaetano Pesce

Figure 3.3　Studio 65 Marilyn, 1972 Stretch fabric-covered moulded polyurethane foam

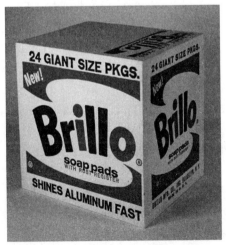

Figure 3.4　Andy Warhol Boxes, 1964 Silkscreen on wood "Brillo"

Figure 3.5　"Just what is it that makes today's home so different, so appealing?" 1956. English artist Richard Hamilton became the father of pop art with this early collage, offering a critical and ironical commentary on the world of conspicuous consumption

Key Words

[1] Pop Design 波普(风格)设计，又称"流行(风格)设计"，是20世纪60年代西方设计界追求形式上的异化及娱乐化的表现主义倾向。波普设计追求大众化的、通俗的趣味、反对现代主义自命不凡的清高，在设计上强调新颖与独特、并大胆采用艳俗的色彩，对20世纪60年代的设计界产生强烈的震动，并对后来的后现代主义产生重要的影响。(注意"Pop Design"波普(风格)设计与"Op Art"欧普艺术的区别，欧普艺术是抽象派绘画风格。)Pop 即 popular 的缩写形式。

[2] Independent Group "独立组"，20世纪50年代英国一个前卫的艺术组织

[3] Richard Hamilton 理查德·汉密尔顿，英国波普艺术重要的代表人物之一

[4] growth [grəʊθ] n. 产品

[5] low art "小艺术"，"低级艺术"。传统上指过去的"工艺美术"或今天的"设计"，与"美术"(包括绘画，雕塑，建筑等)比较而言是"小艺术"，"粗俗的艺术"

[6] planned obsolescence: 有计划的商品废止制，是指产品应被设计和生产满足流行即可，有限时间内使用性和功能性流行于20世纪50年代，由美国工业设计师布鲁克斯·史蒂文提出。

[7] use-it-today, sling-it-tomorrow 用之即弃

[8] polka-dotted a. 带有远点图案的

[9] PVC Blow chair PVC 吹塑椅，PVC 即 polyvinyl chloride "聚氯乙烯"的缩写。

[10] disposable [dɪˈspəʊzəbəl] adj. 1. 一次使用后即丢掉的，一次性的 2. (纳税后的钱)可自由支配的

[11] ephemerality [iˌfeməˈrælətɪ] n. 该词为"物无常用"之意。

[12] Arti-Design association 反设计联盟

[13] Art Nouveau, Art Deco, Futurism, Surrealism, Op Art, Psychedelia, Eastern Mysticism, Kitsch and the Space-Age 新艺术运动、艺术装饰风格、未来派、超现实主义、欧普艺术、视幻艺术、东方神秘主义、通俗文学和太空时代。

Key Sentences

1. The term Pop was coined in the 1950s and referred to the emergence of popular culture during that decade.

"波普"这个词发明于20世纪50年代，用于指在那十年中流行文化的兴起。

2. Pop Design with its Anti-Design association countered the Modern Movement's sober dictum "Less is More" and led directed to the Radical Design of the 1970s. It drew inspiration from a wide range of sources—Art Nouveau, Art Deco, Futurism, Surrealism, Op Art, Psychedelia, Eastern Mysticism, Kitsch and the Space-Age—and was spurred on by the growth of the global mass-media.

波普设计与其反设计联盟反对现代运动的冷静宣言"少就是多"，并且直接导致了20世纪70年代的激进设计。它从新艺术运动、艺术装饰风格、未来派、超现实主义、欧普艺术、视幻艺术、东方神秘主义、通俗文学和太空时代等广泛的来源中获取灵感，并且受到全球大众媒介增长的激励。

课文翻译

波普设计

"波普"这个词语产生于20世纪50年代,指的是那十年期间出现的一种流行文化。1952年,独立组在伦敦成立,它的成员包括画家理查德·汉密尔顿(1922年生)、雕刻家爱德华多·包洛齐(1924年生)、设计评论家雷纳·班汉姆(1920—1988)和建筑师皮特和阿里萨·斯密思,他们是美国第一批探索和歌颂流行消费文化增长的人。

20世纪60年代,美国画家如安迪·沃霍尔(1928—1987)、罗伊·李奇登斯坦(1923—1998)和克拉斯·欧登伯格(1929年生)开始从当代生活各方面的艺术,如广告、包装、漫画和电视中汲取灵感。

不足为奇,波普风格也开始体现在日常用品的设计中,与20世纪50年代的优秀设计相比,设计师更加寻求年轻人的支持,更少地使用庄重的手法。在20世纪50年代,以提高生产力的有计划的商品废止制为代表,产品主导风格为"用之即弃"的思想提供了良好的生长环境,使得这种思想贯穿于20世纪60年代的工业生产中。

彼得·默多克的带有圆点图案的硬纸板"Spotty"婴儿椅(1963年)和帕斯·乌日比诺·洛马齐工作室的PVC吹塑椅是一次性和物无常用这个普遍文化的典型代表。过多的短暂物品,如纸衣服也是如此,它们因风格新颖,使用大量的色彩元素而受到追捧,时尚杂志也变得更加依赖这些物品特征。

由于许多设计师以波普理念进行设计,塑料成为了他们的选择材料。到20世纪60年代,许多新型的塑料和对准工艺可以使用,如注塑成型,并相对便宜。明亮的五彩缤纷的颜色和借助于波普设计辅助的大胆样式席卷战后仅存的节俭风气,反映了20世纪60年代空前的经济繁荣和性解放所带来的普遍的乐观主义思想。自从波普设计瞄准年轻人的市场,产品不得不便宜,从而经常导致低质量。然而,这类产品的可消费性成为其吸引力的一部分,因为它们代表了20世纪50年代倡导的"永恒的"现代经典的反例。

波普设计与其反设计联盟反对现代运动的冷静宣言"少就是多",并且直接导致了20世纪70年代的激进设计。它从新艺术运动、艺术装饰风格、未来派、超现实主义、欧普艺术、视幻艺术、东方神秘主义、通俗文学和太空时代等广泛的来源中获取灵感,并且受到全球大众媒介增长的激励。

但是,20世纪70年代早期的石油危机需要一种更理性的设计方法,于是,波普设计一方面被手工艺复兴运动所取代,另一方面被高科技设计所取代。如果询问起优秀设计的规则以及现代主义的规则,那么,波普设计的影响无疑是深远的,它为后现代主义的产生奠定了一定的基础。

Free Reading 1

Anti-Design

This movement originated in Italy in the 1960s as a reaction against what many avant-garde designers saw as the impoverished language of Modernism, the emphasis placed on style and the aesthetics of good form by many leading manufacturers and celebrated designers. This sense of dissatisfaction with the increasingly widespread diminution of the social relevance of design at the expense of capitalist enterprise had been increasingly aired during the 1950s, particularly in the context of the Milan Triennali. It was also mirrored in the wider economic, political, social, and cultural debates that gripped Italy in the 1960s.

Ettore Sottsass Jr. was a key exponent of the Anti-Design outlook, as were the Radical Design groups Archigram and Superstudio, all expressing their ideas in the production of furniture prototypes, exhibition pieces, and publication of manifestos. Anti-Design sought to harness the social and cultural potential of design rather than embrace style as a means of increasing sales. Where Modernism was typified by notions of permanence, Anti-Design embraced the ephemerality of Pop (shown at the Venice Biennale of 1964), consumerism, and the language of the mass media; where the Modernist palette was generally muted with a prevalence of blacks, whites, and greys, Anti-Design explored the rich potential of colour. Where Modernism admired the integrity of material properties in their own right, Anti-Design embraced ornament and decoration. Furthermore, where Modernism inclined to concepts of Good Design and the adage "form follows function", Anti-Design considered the expressive potential of kitsch, irony, and distortion of scale, characteristics that were also to become hallmarks of Postmodernism and important features of Memphis design.

Free Reading 2

Memphis

The Memphis-Milano Movement was an Italian design and architecture group started by Ettore Sottsass that designed Post Modern furniture, fabrics, ceramics, glass and metal objects from 1981—1987.

The Memphis Group

The Memphis group comprised of Italian designers and architects who created a series of highly influential products in the 1980's. They disagreed with the conformist approach at the time and challenged the idea that products had to follow conventional shapes, colours, textures and patterns.

The Memphis group was founded in 1981. One of the leading members of the group Ettore Sottsass called Memphis design the "New International Style".

Memphis was a reaction against the slick, black humorless design of the 1970's. It was a time of minimalism with such products as typewriters, buildings, cameras, cars and furniture all seeming to lack personality and individualism.

In contrast the Memphis Group offered bright, colourful, shocking pieces. The colours they used contrasted the dark blacks and browns of European furniture. It may look dated today but at the time it looked remarkable. The word tasteful is not normally associated with products generated by the Memphis Group but they were certainly ground breaking at the time.

All this would seem to suggest that the Memphis Group was very superficial but that was far from the truth. Their main aim was to reinvigorate the Radical Design movement. The group intended to develop a new creative approach to design.

On the 11th of December 1980 Scottsass organized a meeting with other such famous designers. They decided to form a design collaborative. It would be named Memphis after the Bob Dylan song "Stuck Inside of Mobile with the Memphis Blues Again". Coincidentally the song had been played repeatedly throughout the evening.

Memphis was historically the ancient Egyptian capital of culture and the birthplace of Elvis Presley. This was quite ironic but so were most of the pieces created by the group.

The image below is of the "Super lamp" created by Martine Bedine. It is made of metal, which has been painted and lacquered. (shown in Figure 3.6)

The group decided that they would meet again in February 1981. By that time each member would have had time to generate design proposals. When they did meet themembers of the group had produced over a hundred drawings, each bold, colourful.

They drew inspiration from such movements as Art Deco and Pop Art, styles such as the 1950's Kitsch and futuristic themes. Their concepts were in stark contrast to so called "Good Design".

The group approached furniture and ceramic companies commissioning them to batch produce their design concepts. On the 18th of September 1981 the group showed its work for the first time at the Arc 74 showroom in Milan. The show exhibited clocks, lighting, furniture and ceramics created by internationally famous architects and designers.

The image below shows the "Carlton bookcase" for Memphis designed by Ettore Sottsass. (shown in Figure 3.7)

Figure 3.6　Martine Bedin: 1980's Memphis Design Modern Super Lamp

Figure 3.7　Carlton bookcase (1981) —Ettore Sottsass

In the same year the group published the book *Memphis, the New International Style*. The book served to advertise the groups work.

Many of the pieces featured in the exhibition were coated in brightly, colourful laminates. Laminates are most commonly used to protect kitchen furniture and surfaces from staining as a result of spillage. The group specifically chose this material because of its obvious "lack of culture".

The work of the Memphis Group has been described as vibrant, eccentric and ornamental. It was conceived by the group to be a "fad", which like all fashions would very quickly come to an end. In 1988 Sottsass dismantled the group.

The group may no longer exist but it has certainly influenced graphic design, restaurant design, fabrics and furnishing.

Free Reading 3

From Ironbridge to the Challenger Space Craft—A Short History of High Tech

Where did High Tech architecture come from? There are two useful historical perspectives, of long range and short range, of 200 years and 20 years. For the long-range perspective, we have to go back to 1779 and the construction of the first cast iron bridge over the River Severn at Coalbrookdale. It is an all-metal prefabricated structure, completely honest in its use of materials and structural forms, but designed as much for elegance as for practicality. In the long term, this must be the favourite candidate for the title "first High Tech structure".

This may seem like far too remote a source for an architectural style born in the 1960s, but the bridge is still standing and we should not underestimate the influence of eighteenth- and nineteenth-century engineering structures on British architects. Decimus Burton's Palm House at Kew Gardens of 1848, the long-span iron, steel and glass roofs over the great railway termini built throughout the second half of the nineteenth century, Eiffel tower and Contamin and Dutert's Galerie des Machines built for the Paris Exhibition of 1889, and of course Paxton's legendary Crystal Palace built for the Great Exhibition of 1851—structures such as these are enduring influences on today's High Tech architects.

They represent an alternative mode of building, based on industrial technology rather than architectural tradition. High Tech architecture shares their confidence and optimism and also, to a large extent, their relatively primitive technology. Throughout the first half of the twentieth century it was to remain an alternative rather than a mainstream mode of building. The characteristic material of Modern-movement mainstream is reinforced concrete, exactly the sort of wet, in situ material that High Tech architects prefer to avoid. Mies van der Rohe is, of course, the exception, but building technology was never his primary concern. The most famous of Mies's construction details—the decorative steel pilasters on the Seagram building—has a dishonesty that most true High Tech architects would deplore. Nevertheless certain habits, the use of external structure, for example, can be traced bock to Mies.

Except in structures that we think of as "pure engineering", the alternative

Modernism was kept alive mainly in theoretical projects, most notably those of the Italian Futurists and the Russian Constructivists. The perspective sketches of Sant' Elia's Citta Nuova, exhibited in 1914, are among the earliest depictions of an architecture that glorifies the technology of concrete, steel, and glass, and which gives dramatic external expression to lift towers, girder bridges, and elevated walkways. The similarities to the more sculpturesque examples of the High Tech style, especially the work of Richard Rogers, is striking. "We no longer believe in the monumental, the heavy and static, and have enriched our sensibilities with a taste for lightness, transience and practicality," wrote Sant' Elia in the catalogue to the Citta Nuova exhibition. "We must invent and rebuild ex novo our modern city like an immense and tumultuous shipyard, active, mobile and everywhere dynamic, and the modern building like a gigantic machine. Lifts must not longer hide away like solitary worms in the stairwells... but must swarm up the facades like serpents of glass and iron." The Centre Pompidou and the Lloyd's building would be quite at home in the Citta Nuova. With the Russian Constructivists we come even closer to the precise sensibilities of High Tech. Look, for example, at Iakov Chernikhov's "Fantasies", the Constructivist equivalent of Sant' Elia's visionary drawings; or at Alexander Vesnin's project for the Pravda building in Moscow of 1923. This bristles with proto-High Tech motifs, such as diagonal steel cross-bracing, lifts in glass shafts and even what appears to be a satellite dish on the roof (in fact it is a searchlight). We can even begin, at this point, to trace direct and acknowledged influences on High Tech. In Western Europe, the influence of Constructivism was felt most strongly in the Netherlands and is most visible in the work of Mart Stam, who collaborated with El Lissitsky, the chief propagandist of Constructivism, and Johannes Duiker. Duiker's partner, Bernard Bijvoet, was to collaborate with Pierre Chareau in the design of the Maison de Verre in Paris, completed in 1932. This building is a curious assemblage of mass-produced, machine-like components with a flexible plan and an external wall made entirely of glass lenses. In 1959 Richard Rogers visited the Maison de Verre and he now acknowledges it as the building that has had the most influence on his architecture.

While Chareau and Bijvoet were designing the Maison de Yerre, Jean Prouvé was developing the first system af replaceable wall components for lightweight metal houses. Prouvé was to continue to develop his own, peculiarly French, metal and glass architecture right up to the 1970s. The extent of his influence on British High Tech can be gauged by Norman Foster's remark when he invited Prouvé to visit his office: "We would never have done all this without you."

Meanwhile, on the other side of the Atlantic, Buckminster Fuller was proposing an even more thoroughgoing application of advanced technology in his Dymaxion House project of 1927, a hexagonal structure of lightweight metal and plastic suspended from a core of mechanical services. If anyone deserves the title "father of High Tech" it is Fuller. His comprehensive and knowledgeable use of materials and technology borrowed from other industries (the Dymaxion House proposed an adaptation of techniques used in aircraft construction at that time), his insistence on a global view of building performance (architects, he said, should know not just how big their buildings were, or how much they cost but also how much they weighed), and his refusal to have anything to do with the conventions of traditional, academic architecture—these have all been built into the ideological structure of High Tech.

It was Reyner Banham, in the closing pages of *Theory and Design in the First Machine Age*, who first introduced Fuller to British architects as a possible model for the future development or modern architecture. This was in 1960 and coincided with an outpouring of formally and technologically inventive projects from students and teachers at London schools of architecture. Especially the Regent Street Polytechnic and the Architectural Association, The group called Archigram (Peter Cook, Warren Chalk, David Greene, Denis Crompton, Ron Herron and Mike Webb) began to publish and exhibit spectacular theoretical projects that clearly displayed many of the features of the High Tech architecture of the 1970s and 1980s: the indeterminate forms, the mass-produced, expendable, plug-in components, the use of technologies from the emerging aerospace industry and, above all, the idea that architects had a duty to increase personal, environmental choice. Architectural historian Robin Middleton has remarked that in the 1960s Archigram did for architecture what the Beatles did for music. Richard Rogers, Nicholas Grimshaw, and Michael Hopkins were all students at the AA. Of the big four High Tech architects, only Norman foster, who studied at Liverpool school of architecture, was not directly exposed to the influence of the AA in the early 1960s, There is no doubt that projects such as Michael Webb's "bowellist" Furniture Manufacturer's Association Headquarters, a student project of 1958, Cedric Price's Fun Palace of 1963, Peter Cook's Plug-in City of 1964, and Ron Herron's Walking City of the same year were well known to Rogers, Foster and Co. They were, after all, well known much further afield, largely due to the publicizing efforts of Reyner Banham. The capsule buildings of the Japanese Metabolists, for example, surely owe a debt to Cook's Plug-in City. But we must not make the mistake of assuming that High Tech is simply built Archigram. There were other, and possibly more important, contemporary influences, both British and American. Of the British influences, Alison and Peter Smithson and James Stirling are the most important. All were teachers at the Architectural Association at the relevant time. The Smithsons' Hunstanton School, designed in 1949 and sometimes described as "Miesian brutalist", was one of the very few British postwar buildings to be accorded any respect by the 1950s avant garde. It had a curiously formal, Palladian plan, but what made it revolutionary at the time was the way it displayed with complete honesty its materials—steel frame, brick infill, precast concrete floors, exposed electrical conduit and pipework, and a proprietary steel water tank raised on a freestanding steel frame like a campanile. James Stirling's Engineering Building at Leicester University, designed in partnership with James Gowan in 1959 and completed in 1963, is another historical marker in the development of British Modernism. A powerful composition in red tile-clad concrete and patent glazing, it combined constructivism, nineteenth century engineering, and the colours and textures of red-brick city of Leicester in such an utterly convincing way that it made James Stirling's international reputation almost overnight.

Both of these buildings can be seen, with mindsight, to contain the seeds of High Tech and both were, without question, powerful influences during High Tech's "student years". The influence of Stirling, in particular, must be emphasized. If it were not for his subsequent metamorphosis into Britain's leading Post Modernist, and therefore at the opposite stylistic pole from High Tech, we might now be describing his 1964 History Faculty library at Cambridge University as the first British High Tech building. It has so many of the High Tech motifs: a glass skin, a freestanding lift and service tower, a prominent roof-mounted maintenance crane, huge lattice trusses over the reading room and, most telling, three air extract units slung between the trusses, clearly visible from below and painted in primary colours.

Rogers, Foster and Stirling all come together for a brief period in 1962 at Yale University where Rogers and Foster were postgraduate students, collaborating for the first time, and Stirling was a visiting critic. The influence of American architects such as Paul Rudolph (then chairman of the architecture school at Yale) and Louis Kahn are detectable in some of the later High Tech buildings. Kahn's concept of "served" and "servant" spaces is especially important: compare the satellite servant towers of Rogers' Lloyd's building with the brick-clad service towers of Kahn's Richards Medical Research Building in Philadelphia, of 1961. But the strongest American influence was Californian—the simple, flexible, lightweight steel and glass houses of Charles Eames, Craig Ellwood and Raphael Soriano. These were illustrated in a 1962 book by Esther McCoy called Modern California Houses (republished in 1978 as Case Study Houses: 1945-1962), which was to be a source of inspiration for Rogers and Foster when they returned to England and set up in practice together under the name Team 4.

Team 4's first important building, a house at Creek Vean in Cornwall, could not remotely be described as High Tech. Concrete blockwork was its main material and its main influence was not Archigram or Eames but Frank Lloyd Wright. The only advanced technical feature was the Neoprene used to seal the sloping glazing. (The Neoprene gasket was to become one of the distinguishing marks of High Tech in its formative years.)

In the short-range, 20-year perspective of the history of High Tech, the title "first British High Tech building" must go to the simple, single-storey Reliance Controls Electronics Factory of 1967 at Swindon. Ironically this was the last building on which Rogers and Foster collaborated. It was Miesian in conception and owed a lot to the much larger Cummins Engine Company factory at Darlington, designed by the American firm of Kevin Roche, John Dinkeloo and Associates and completed in 1965. It had a simple rectangular plan, a fiat roof, and a freestanding water tower copied from the Smithsons' Hunstanton School. It would have been quite unremarkable were it not for the exposed steel structure (painted white), the flexible and extendable multi-purpose plan, and the way that it was rapidly assembled from dry, off-the-shelf components. This was the first of a long line of simple, elegant factory/office buildings designed by High Tech architects for High Tech (in the industrial sense) clients. It was instantly acclaimed by critics, won the Financial Times award for the most outstanding industrial building of 1967, and gave its designers the confidence to develop the new style with renewed energy.

A close rival for the title "first British High Tech building" is the glass-clad spiral of plastic bathroom pods designed in 1967 by Nicholas Grimshaw to plug into the rear of a Victorian house being converted into a student hostel by his then partner Terry Farrell. In Reliance Controls, the mechanical services had been hidden in a floor duct, so it was Grimshaw who first realized the plug-in servant tower concept that was to become a prominent High Tech motif in subsequent years.

The High Tech repertoire was now complete and for the next ten years each element was developed with ever-increasing confidence in successive buildings and projects by Rogers, Grimshaw, Foster and Michael Hopkins, who joined Foster's office in 1969. There was the slick glass and Neoprene skin of Foster's Amenity Centre for the Fred Olsen shipping line in the London Docks, the severely minimal single-storey office building for IBM at Cosham, designed by Foster and Hopkins in 1971, Farrell and Grimshaw's factories for Herman Miller, and a series of vehicle-like buildings and projects by Rogers, with highly insulated, zip-up skins and round-cornered windows fixed, once again, with Neoprene gaskets.

In 1970 Rogers entered into partnership with Renzo Piano, who had been developing his own, highly sophisticated Italian version of High Tech in his office in Genoa. In 1971 the new partnership entered and won the international competition for the design of a new national art and culture centre on the Plateau Beaubourg in Paris. With the Centre Pompidou, High Tech came of age. Pompidou had everything: flexible plan, exposed structure, plug-in services, and the glorification of machine technology. When it was completed in 1977, the image of High Tech suddenly came into focus, entered the public consciousness, and became an internationally influential style. Some of the best examples of that style, mainly from the decade after Pompidou, are illustrated in this book.

So the 20-year story of High Tech has a beginning, the Reliance Controls Factory, a middle, the Centre Pompidou and, just possibly, an end in the two masterpieces completed in 1986, the Hongkong Bank Headquarters by Norman Foster and Lloyd's of London by Richard Rogers. For there are signs that High Tech is running out of steam, The latest Rogers and Foster projects demonstrate a diminishing interest in technology and a new concern for the less tangible aspects of the complex relationships between people and spaces, and between buildings and cities.

Foster's design for a Mediatheque in the centre of Nîmes, opposite the Roman Temple known as the Maison Carrée, proposes a very un-High Tech palette of materials: concrete, bronze and local stone. On an early published sketch there appears the following note, "No diagonals in structure—must not look industrial" And Foster's abortive scheme for the new headquarters for BBC Radio in London is curiously subdued—a humble exercise in urban contextualism rather than a glorification of technology. Rogers, too, seems to have discovered the existing city as an architectural theme. When asked to provide a theoretical project for the 1986 Foster, Rogers, Stirling exhibition at the Royal Academy, he presented a scheme for the revitalization of London's South Bank. The scheme included a bombastic and highly technological new bridge across the Thames, but its main thrust was an almost Baroque realignment of vistas and reinforcement of public spaces.

But perhaps the most important change has been in the technological, rather than the architectural, climate. Technology has moved on and once again left architecture behind. There may be an architectural equivalent of the aeroplane or the lunar module, but there is no architectural equivalent of the silicon chip. The aerospace industry has always been the happy image-hunting ground of the High Tech architect but it no longer holds the fascination and promise that it did in the late 1960s and 1970s. Architectural scholiasts of the future, wishing to pin down the precise date of the death of the High Tech style, might well choose January 28th 1986, the day the Challenger space craft blew up in front of the watching millions. The cause of the tragedy, we now know, was the failure of a Neoprene gasket.

Free Reading 4

The Post-Standardized Mind

The Third Wave does more than alter Second Wave patterns of synchronization. It attacks another basic feature of industrial life: standardization.

The hidden code of Second Wave society encouraged a steamroller standardization of many things—from values, weights, distances, sizes, time, and currencies to products and prices. Second Wave businessmen worked hard to make every widget identical, and some still do.

Today's wisest businessmen, as we have seen, know how to customize (as opposed to standardize) at lowest cost, and find ingenious ways of applying the latest technology to the individualization of products and services. In employment the numbers of workers doing identical work grows smaller and smaller as the variety of occupations increases. Wages and fringe benefits begin to vary more from worker to worker. Workers themselves become more different from one another, and since they (and we) are also consumers, the differences immediately translate into the marketplace.

The shift away from traditional mass production thus is accompanied by a parallel de-massification of marketing, merchandising, and of consumption. Consumers begin to make their choices not only because a product fulfills a specific material or psychological function but also because of the way it fits into the larger configuration of products and services they require. These highly individualized configurations are transient, as are the life-styles they help to define. Consumption, like production, becomes configurational post-standardized production brings with it post-standardized consumption.

Even prices, standardized during the Second Wave period, begin to be less standard now, since custom products require custom pricing. The price tag for an automobile depends on the particular package of options selected; the price of a hi-fi set similarly depends on the units that are plugged together and on how much work the buyer wishes to do; the prices of aircraft, offshore oil rigs, ships, computers, and other high-technology items vary from one unit to the next.

In politics we see similar trends. Our views are increasingly nonstandard as consensus breaks down in nation after nation and thousands of "issue groups" spring up, each fighting for its own narrow, often temporary, set of goals. In turn, the culture itself is increasingly de-standardized.

Thus we see the breakup of the mass mind as the new communications media come into play. The de-massification of the mass media—the rise of mini-magazines, newsletters, and small scale, often Xeroxed, communications along with the coming of cable, cassette, and computer—shatters the standardized image of the world propagated by Second Wave communications technologies, and pumps a diversity of images, ideas, symbols, and values into society. Not only are we using customized products, we are using diverse symbols to customize our view of the world.

Art News summarized the views of Dieter Honisch, director of the National Gallery in West Berlin, "What is admired in Cologne may not be accepted in Munich and a Stuttgart success may not impress the Hamburg public. Ruled by sectional interests, the country is losing its sense of national culture."

Nothing underlines this process of cultural de-standardization more crisply than a recent article in Christianity Today, a leading voice of conservative Protestantism in America. The editor writes, "Many Christians seem confused by the availability of so many different translations of the Bible. Older Christians did not face so many choices." Then conies the punch line, "Christianity Today recommends that no version should be the 'standard.'" Even within the narrow bounds of Biblical translation, as in religion generally, the notion of a single standard is passing. Our religious views, like our tastes, are becoming less uniform and standardized.

The net effect is to carry us away from the Huxleyan or Orwellian society of faceless, de-individualized humanoids that a simple extension of Second Wave tendencies would suggest and, instead, toward a profusion of life-styles and more highly individualized personalities. We are watching the rise of a "post-standardized mind" and a "post-standardized public".

This will bring its own social, psychological, and philosophical problems, some of which we are already feeling in the loneliness and social isolation around us, but these are dramatically different from the problems of mass conformity that exercised us during the industrial age.

Because the Third Wave is not yet dominant even in the most technically advanced nations, we continue to feel the tug of powerful Second Wave currents. We are still completing some of the unfinished business of the Second Wave. For example, hard-cover book publishing in the United States, long a backward industry, is only now reaching the stage of mass-merchandising that paperback publishing and most other consumer industries attained more than a generation ago. Other Second Wave movements seem almost quixotic, like the one that urges us at this late stage to adopt the metric system in the United States to bring American measurements into conformity with those used in Europe. Still others derive from bureaucratic empire building, like the effort of Common Market technocrats in Brussels to "harmonize" everything from auto mirrors to college diplomas— "harmonisation" being the current gobbledygook for industrial-style standardization.

Finally, there are movements aimed at literally turning back the clock—like the back-to-basics movement in United States schools. Legitimately outraged by the disaster in mass education, it does not recognize that a de-massified society calls for new educational strategies, but seeks instead to restore and enforce Second Wave uniformity in the schools.

Nevertheless, all these attempts to achieve uniformity are essentially the rearguard actions of a spent civilization. The thrust of Third Wave change is toward increased diversity, not toward the further standardization of life. And this is just as true of ideas, political convictions, sexual proclivities, educational methods, eating habits, religious views, ethnic attitudes, musical taste, fashions, and family forms as it is of automated production. An historic turning point has been reached, and standardization, another of the ruling principles of Second Wave civilization, is being replaced.

Unit Four
Contemporary Industrial Design Theory and Practice

Core Text 10

Industrial Design

Design process applied to goods produced, usually by machine, by a system of Mass Production based on the division of labour. The terms "industrial design" and "industrial designer" were first coined in the 1920s in the USA to describe those specialist designers who worked on what became known as product design. The history of industrial design is usually taken to start with the industrialization of Western Europe, in particular with the Industrial Revolution that began in Britain in the second half of the 18th century. The main focus of attention then moved to the USA and Germany when they industrialized and, in the second half of the 19th century, began to challenge Britain's supremacy as "the workshop of the world". After World War II, Italy, the former Federal Republic of Germany and Japan each challenged the USA for world supremacy in industrial design.

When the industrial design profession was becoming firmly established in American manufacturing industry, *Industrial Designhas* long been established as the leading magazine for industrial designers in the United States. Retitled *ID* Magazine of Industrial Design in 1980, since its establishment it has carried a broad range of critical material relating to the practice, culture, and business of design. It is published eight times a year, including the prestigious *Annual Design Review* that embraces consumer products, furniture, equipment, environments, packaging, graphics, and student projects. Some industrial product are shown as follows: (See Figure 4.1—Figure 4.5)

Figure 4.1 An iPod, an industrially designed product

Figure 4.2 Kitchen Aid 5 qt. Stand Mixer, a design hardly changed since its original introduction in the 1930s and still successful today

Figure 4.3 Western Electric model 302 Telephone, found almost universally in the United States from 1937 until the introduction of touch-tone dialing

Figure 4.4 A Fender Stratocaster with sunburst finish, one of the most widely recognized electric guitars in the world

Figure 4.5 Model 1300 Volkswagen Beetle

Definition of Industrial Design

General Industrial Designers are a cross between an engineer and an artist. They study both function and form, and the connection between product and the user. They do not design the gears or motors that make machines move, or the circuits that control the movement, but they can affect technical aspects through usability design and form relationships. And usually, they partner with engineers and marketers, to identify and fulfill needs, wants and expectations.

In depth "Industrial Design (ID) is the professional service of creating and developing concepts and specifications that optimize the function, value and appearance of products and systems for the mutual benefit of both user and manufacturer" according to the IDSA (Industrial Designers Society of America).

Design, itself, is often difficult to define to non-designers because the meaning accepted by the design community is not one made of words. Instead, the definition is created as a result of acquiring a critical framework for the analysis and creation of artifacts. One of the many accepted (but intentionally unspecific) definitions of design originates from Carnegie Mellon's School of Design, "Design is the process of taking something from its existing state and moving it to a preferred state."

Process of Design

Although the process of design may be considered "creative", many analytical processes also take place. In fact, many industrial designers often use various design methodologies in their creative process. Some of the processes that are commonly used are user research, sketching, comparative product research, model making, prototyping and testing. These processes can be chronological, or as best defined by the designers and/or other team members. Industrial Designers often utilize 3D software, Computer-aided industrial design and CAD programs to move from concept to production.

Product characteristics specified by the industrial designer may include the overall form of the object, the location of details with respect to one another, colors, texture, sounds, and aspects concerning the use of the product ergonomics. Additionally the industrial designer may specify aspects concerning the production process, choice of materials and the way the product is presented to the consumer at the point of sale. The use of industrial designers in a product development process may lead to added values by improved usability, lowered production costs and more appealing products. However, some classic industrial designs are considered as much works of art as works of engineering: the iPod, the Jeep, the Fender Stratocaster, the Coke bottle, and the VW Beetle are frequently-cited examples.

Industrial design also has a focus on technical concepts, products and processes. In addition to considering aesthetics, usability, and ergonomics, it can also encompass the engineering of objects, usefulness as well as usability, market placement, and other concerns such as seduction, psychology, desire, and the emotional attachment of the user to the object.

Product design and industrial design can overlap into the fields of user interface design, information design and interaction design. Various schools of industrial design and/or product design may specialize in one of these aspects, ranging from pure art colleges (product styling) to mixed programs of engineering and design, to related disciplines like exhibit design and interior design, to schools where aesthetic design is almost completely subordinated to concerns of function and ergonomics of use (the so-called *functionalist* school).

Key Words

[1] industrial design 工业设计

[2] supremacy [suːˈpreməsɪ] *n.* 至高无上；最高权力

[3] packaging [ˈpækɪdʒɪŋ] *n.* 包装材料；包装材料的设计和生产

[4] graphics ['græfɪks] n. 图样；图案；绘图；图像
[5] usability design 可用性设计
[6] artifact ['ɑːtəfækt] n. 人工制品；手工艺品；加工品
[7] sketching ['sketʃɪŋ] n. 草图
[8] chronological [krɒnə'lɒdʒɪkəl] adj. 1. 按时间的前后顺序排列的 2. 按时间计算的（年龄）（相对于身体、智力或情感等方面的发展而言）adv. 按时间的前后顺序排列的
[9] computer-aided industrial design 计算机辅助工业设计
[10] ergonomics [ˌɜːɡə'nɒmɪks] n. 人类工程学，生物工程学
[11] esthetics [iːs'θetɪks] n. 美学，美术理论，审美学，美的哲学
[12] exhibit design 展示设计
[13] interior design 室内设计
[14] subordinate [sə'bɔːdɪnət]: adj. 1. 级别或职位较低的，下级的 2. 次要的，附属的 vt. 从属

Key Sentences

1. General Industrial Designers are a cross between an engineer and an artist. They study both function and form, and the connection between product and the user. They do not design the gears or motors that make machines move, or the circuits that control the movement, but they can affect technical aspects through usability design and form relationships. And usually, they partner with engineers and marketers, to identify and fulfill needs, wants and expectations.

通常，工业设计师是工程师与艺术家的结合。他们研究功能和形态，也研究产品和用户的关联。他们并不设计使机器运转的齿轮和马达，也不设计电路以控制运动，但他们可以通过可用性设计和结构关系对技术方面产生影响。通常，他们与工程师和市场营销人员合作，以确定和满足需求、需要和期望。

2. In depth "Industrial Design (ID) is the professional service of creating and developing concepts and specifications that optimize the function, value and appearance of products and systems for the mutual benefit of both user and manufacturer" according to the IDSA (Industrial Designers Society of America).

深入地讲，根据美国工业设计师协会（IDSA）的定义，"工业设计是提供创造和发展概念的专业化服务，优化其功能、价值和产品的外观，是用户和制造商互利互惠的系统"。

课文翻译

工 业 设 计

设计过程通常通过机械和基于劳动分工的批量生产系统而应用于商品生产。"工业设计"和"工业设计师"的术语首次用于20世纪20年代的美国，被用来描述那些因为产品设计而著名的专家设计师们。工业设计的历史通常开始于西欧工业化，特别是18世纪下半叶始于英国的工业革命。19世纪下半叶，人们所关注的主要焦点则转移到美国和德国，

当时它们已经完成工业化，并开始挑战英国的"世界工厂"的霸主地位。第二次世界大战后，意大利、前德意志联邦共和国和日本各自挑战美国在工业设计领域的世界霸权地位。

当工业设计职业首先在美国制造业中产生之时，《工业设计》作为美国工业设计师的领导性杂志成立。1980年，它被重命名为《工业设计》杂志，从其创立至今，已经刊登了广泛的重要性题材，其中涉及实践、文化和设计商务。该杂志每年出版8次，其中包括著名的设计年度回顾、消费产品、家具、设备、环境、包装、图形以及学生作品。

工业设计的定义

通常，工业设计师是工程师与艺术家的结合。他们研究功能和形态，也研究产品和用户的关联。他们并不设计使机器运转的齿轮和马达，也不设计电路以控制运动，但他们可以通过可用性设计和结构关系对技术方面产生影响。通常，他们与工程师和市场营销人员合作，以确定和满足需求、需要和期望。

深入地讲，根据美国工业设计师协会（IDSA）的定义，"工业设计是提供创造和发展概念的专业化服务，尽可能优化其功能、价值和产品的外观，是用户和制造商互利互惠的系统"。

设计本身对于非设计师往往很难定义，因为，设计团体所认同的设计定义不只是这个词的片面意义。相反，这个定义的创建是由于人工产品的分析和创造而获得重要构架的结果。众多被接受的设计定义之一来自于卡内基·梅隆设计学院的说法，"设计是将其目前状态提升到一个更好状态的过程"。

设计的过程

虽然设计过程被认为是创造性的，但也产生了许多分析过程。事实上，许多工业设计师经常在他们的创新过程使用多种设计方法。经常用到的一些设计过程包括用户研究、草图、产品比较研究、模型制作、原型和测试。这些过程可以按时间顺序排列，或者由设计师或者其他团队成员做最佳界定。工业设计师通常使用三维软件、计算机辅助工业设计与计算机辅助设计（CAD）程序等完成从概念到生产的过程。

由工业设计师指定的产品特性可能包括物体的整体形式、细节的相互定位、颜色、质地、声音以及各方面有关产品的人体工程学运用。此外，工业设计师会特别关注生产过程、材料的选择及在销售点产品向消费者展示的方式。工业设计师在产品发展过程中的作用可能通过提高产品使用性而增加其价值，降低生产消耗，使产品更具吸引力。然而，一些经典工业设计的艺术性和工程性同样被深思熟虑，苹果iPod、吉普汽车、芬达电吉他、可乐瓶以及大众甲壳虫汽车都是常常被引用的例子。

工业设计也关注技术概念、产品和过程。除了考虑到美学、使用性和人体工程学，工业设计同时也关注物品的工程性和使用性相同的有用性、市场布局以及其他因素，例如，使用者对物品的关注、心理、需求以及情感依赖。

产品设计和工业设计可以与用户界面设计、信息设计和交互设计领域相重叠。各个学校的工业设计和/或产品设计可能会专攻某一个方向，从纯粹的艺术院校（产品造型）到工程和设计的混合专业，到展览设计和室内设计相关学科，到美学设计学校几乎完全服从于功能和人体工程学的使用问题（即所谓的功能主义学校）。

Free Reading 1

ICSID

About ICSID?

The International Council of Societies of Industrial Design (ICSID) is a non-profit organization that protects and promotes the interests of the profession of industrial design.

Founded in 1957, ICSID serves as a unified voice of over 50 nations through which members can express their views and be heard on an international platform. Since its inception, ICSHID has continued to develop its wide-reaching network of students and professionals devoted to the recognition, success and growth of the industrial design community.

Together, professional associations, promotional societies, educational institutions, government bodies and corporations create a comprehensive and diverse system on the forefront of industrial design education and progress.

Mission & Vision Statements

The primary aim of the association is to advance the discipline of industrial design at an international level. To do this, ICSID undertakes a number of initiatives of global appeal to support the effectiveness of industrial design in an attempt to address the needs and aspirations of people around the world, to improve the quality of life, as well as help to improve the economy of nations throughout the world.

Vision

ICSID strives to create a world where design enhances our social, cultural, economic and environmental quality of life.

Mission

- To facilitate collaboration between the membership pillars of ICSID

Although there are many international organisations that represent one aspect of industrial design, ICSID is the only one that includes member organisations from the professional, promotional, educational and corporate spheres. As a result, ICSID is uniquely positioned to foster communication and connectivity across the spectrum of the profession. ICSID includes all membership pillars in its events, activities and publications and promotes ways that they can collaborate with one another.

- To support and protect the professional practice of industrial design at regional and international levels

ICSID was founded to further and protect the professional practice of industrial design internationally, and this founding principle remains central to our identity today. ICSID is an advocate on behalf of professional designers in such areas as: design standards, the appreciation of professional achievement, fair practices, sustainable development and support of intellectual property rights, from advanced technology to traditional knowledge and crafts.

- To contribute to the advancement of the study of design in theory, research and practice at all levels of education

ICSID is committed to promoting the highest quality of design education and professional preparation through establishing opportunities to study industrial design in all parts of the world, sharing best practices, recognising student achievement, connecting educators to the profession and developing an international education network.

- To collaborate with stakeholders in design and other international organisations to foster a global understanding of design

As the international voice for industrial design, ICSID has the opportunity and responsibility to promote the value and benefit of design around the world. We work with local and national governments, international media, corporations, and non-governmental organisations to foster a global understanding of design and its power to enhance quality of life.

Free Reading 2

IDSA

Industrial Designers Society of America (IDSA) is an organization of professional industrial designers in the United States. IDSA is also a member of the International Council of Societies of Industrial Design (ICSID).

The Industrial Designers Society of America (IDSA) is the world's oldest, largest, member-driven society for product design, industrial design, interaction design, human factors, ergonomics, design research, design management, universal design and related design fields. IDSA organizes the renowned International Design Excellence Award (IDEA) competition annually; hosts the International Design Conference and five regional conferences each year; and publishes *Innovation*, a weekly e-newsletter highlighting the latest headlines in the design world. IDSA's charitable arm, the Design Foundation, supports the dissemination of undergraduate scholarships annually to further industrial design education.

The roots of IDSA can be traced back to a number of earlier organizations set up before the Second World War. These included the *American Union of Decorative Artists and Craftsmen* (AUDAC), founded in 1928, the *National Furniture Designers' Council* (NFDC) which lasted from 1933 to 1934, and the Designers' Institute of the American Furniture Mart, founded in 1936. Two years later the latter formed the Chicago-based *American Designers Institute* (ADI), which drew on a much wider range of specialist expertise, ranging from industrial design to graphics and the decorative arts under the presidency of John Vassos.

In 1951 ADI moved its administration to New York, changed its name to the *Industrial Designers Institute* (IDI), and established an annual programme of National Design Awards which ran until 1965. However, rather more significant was the *Society of Indus-*

trial Designers (SID), which was founded in 1944 by fifteen designers who were well grounded in industrial design practice. The SID became the *American Society of Industrial Designers* (ASID) in 1957, the same year in which the *Industrial Design Education Association* (IDEA) was founded as neither of the professional bodies admitted educators. By 1965 these three bodies had between them a membership of about 600 representing a wide spectrum of professional interests relating to industrial design and they merged to become the Industrial Designers Society of America (IDSA). Its first chair was John Vassos and first president Henry Dreyfuss. In 1980 the annual National Design Awards were re-established. Membership in 2000 amounted to more than 3,000.

IDSA's Core Purpose and Mission

IDSA's core purpose is to advance the profession of industrial design through education, information, community and advocacy.

IDSA's mission is:

1. Lead the profession by expanding our horizons, connectivity and influence, and our service to members.

2. Inspire design quality and responsibility through professional development and education.

3. Elevate the business of design and improve our industry's value.

Free Reading 3

JIDPO

Established in 1969, the Japan Industrial Design Promotion Organization (JIDPO) serves as Japan's only institution with comprehensive advocacy towards the discipline of design. JIDPO contributes to improving quality of life and industrial output through numerous design promotion activities, the pinnacle of which is the Good Design Awards program. In order to foster a broad understanding with respect to the value that design provides and to make design relevant across many lifestyle settings, JIDPO continually strives to increase its level of design advocacy, while supporting the extensive community associated with design and implementing various programs that deepen understanding about design. The Good Design Awards, today practically a public monument to design in Japan, provide a platform for publicity throughout Japan and around the globe. JIDPO continues to develop forums where large numbers of people can partake of the benefits that design brings to our lives.

History of Design Promotion in Japan

Over a half century has transpired since Japan's design promotion activities were systematically organized. The subjects associated with design promotion change through the years. The history, therefore, marks milestones, which indicate the magnitude of need for design in industry and society within Japan. The following traces a brief introduction to this historical progression.

Approximately 150 years ago when Japan's modernization commenced, activities to improve the form and color of chinaware and handicrafts, which constituted the principal exports at the time, began. The effort merely addressed improvements in exterior design. But in 1928, a national laboratory, though small, was established, and the study and practical instruction of design under the modern meaning started. Renowned designers like Bruno Taut and Charlotte Perriand were invited here, and traditional craftsmanship was reassessed from an industrial perspective.

- 1957

Good Design Product Selection Program is launched (known as Good Design Awards today).

- 1958

Design Section is installed at the Ministry of International Trade and Industry (Design Policy Office of Ministry of Economics, Trade, and Industry today).

- 1960

JETRO opens Japan Design House.

Despite a respite from the post-war turmoil, Japan remained a seriously impoverished nation. No policy could lift the nation out of poverty, except growing exports of industrial goods.

Once started, however, exporting revealed the lack of technology for creating competitive products in the market. The leadership of Japan began to realize that the technology concerned design, and an organization was initiated as part of industry promotion policy. The Good Design Awards, which remain a principal program of design promotion today, were instituted, and design instruction and training took place actively under the invitation of designers from the United States.

Trade organizations, such as the Japan Industrial Design Association and Japan Design Committee, were established, and activities of designers began to ramp up.

- 1961

The Council of Design Promotion of MITI proposes establishment of a design promotion institution.

- 1969

Japan Industrial Design Promotion Organization (JIDPO) is established.

- 1971

JIDPO joins ICSID.

- 1973

ICSID convention in Japan results in 1973 Design Year action program in cooperation with JIDA.

As exports flourished, living standards in Japan grew in affluence, and domestic markets developed around demand for durable consumer goods. The power of design in industrial activity gained increasing understanding as a result. An institution with a comprehensive promotion mandate for design came in to demand. Those needs culminated in the es-

tablishment of the Japan Industrial Design Promotion Organization.
- 1974

JIDPO receives the service contract for Good Design Product Selection Program from MITI.
- 1975

JIDPO launches regional design development activities.
- 1981

Japan Design Foundation is established.

JIDPO took over the activities of JETRO's Japan Design House, which was engaged in promoting exports of household goods produced by local industry, and design to major manufacturing businesses. JIDPO also inherited the secretariat of the Good Design Product Selection Program, previously managed by the Japan Chamber of Commerce. Activities and organizations gradually grew in robustness alongside the handling of existing design promotion activities. In particular, the regional design development activities integrated a variety of local industry support measures from the perspective of design. An alliance with nearly half of the nation's prefectures led to a ten-year program that targeted major centers of industry around the nation.

As a dedicated institution to foster international exchange through design, the Japan Design Foundation (JDF) was established. Until its role concluded in 2009, JDF promoted international design competitions and design exchange activities with various Asian nations.
- 1984

Good Design Product Selection Program greatly expands subject domains.
- 1988

The Council of Design Promotion pronounces "Design Policy for the 1990s" and the Design Year.
- 1989

The 1989 Design Year is instituted as a nationwide activity with JIDPO serving as secretariat.

As the quality of life in Japan improved during the 1980s, the Japanese market expanded tremendously, and Japanese industry flourished to unprecedented proportions. Growth prompted the need for design in all corners of industry, while promotion activities to transfer design technology from consumer product domains to industrial product and public utility product domains achieved tremendous results. Maturation of society as a whole led to an understanding about how the power of design could support many kinds of activities for citizens. MITI and JIDPO proposed the activation of design competency to daily living and community development, and called upon local government, leading design firms, and academic institutions of design to engage in new activities through design. This action program held under the Design Year banner grew to a very large scale, and included the World Design Expo, which commemorated 100 years of municipal history of Nagoya and brought the participation of 400 businesses. The action program expanded the domain of design promotion to spheres encompassing citizens and daily living.

- 1993

The Council of Design Promotion releases "New Design Policy to Meet the Changes of Our Times."

JIDPO institutes the Design Resources Development Center.

- 1998

Good Design Product Selection Program is privatized, and JIDPO carries on with the program as the Good Design Awards.

In the 1990s, society experienced structural transformations, and designers found that the industrialized society mode no longer matched the scenario. The Council of Design Promotion recommended that training of new design resources should take place. JIDPO responded by instituting the Design Resources Development Center and initiating new activities. As part of the National Government reforms, MITI relinquished sponsorship of the Good Design Product Selection Program, and the privatized operation re-launched as the Good Design Awards under the sponsorship of JIDPO. These changes prompted JIDPO to pursue a service model, where promotion activities that increased public value were provided as a service business.

- 2004

Good Design Awards encompass the ASEAN Selection.

Tokyo Design Market is launched under contract from the Tokyo Metropolitan Government.

Design Excellence Company Awards are launched.

With the arrival of the 21st century, the structural transformations in society started to become more evident, and expectations grew for the potential held by design in planning. Design gained an appreciation for its resource value in society, and a school of thought emerged, which placed design in the driver's seat for society in general.

Tokyo Design Market fostered a forum for designers with concept-pitching abilities and manufacturers excelling in making products to meet. The intent was to stimulate new products and new business. The Design Excellence Company Awards positioned design as a business resource, and commended business owners who provided products and services of exemplary value for living.

- 2006

Good Design Awards mark 50 years since establishment. JIDPO conducts commemorative activities, including the Milan exhibition.

- 2007

JIDPO relocates head offices to Tokyo Mid-Town, and institutes the Design Liaison Center in an alliance with overseas academic research institutions.

Design Hub activities launch in cooperation with JAGDA, Kyushu University.

The Good Design Awards celebrated their 50th anniversary of establishment in 2006.

During the half-century, over 35,000 entries had received a Good Design Award. Steadily rising recognition over the years brought the program to a point where over 1,500 companies and designers applied annually with over 3,000 entries in total, and awareness for the program reached 87%. The Good Design Awards matured into a social institution as a route to obtain support from consumers, in addition to industry and designers. The numbers demonstrated that design promotion in Japan had achieved certain, measurable success.

Today, expectations for design are surpassing its legacy in industrialized society with the arrival of new questions. What is a sustainable society? What needs specific implementation? Design is being tasked with new answers to these subjects, and we are called to prepare the forums for this problem solving work.

JIDPO, in fact, takes the role of preparing the forums. In this regard, the Design Hub and Liaison Center were instituted at the time of our office relocation to Roppongi. The agenda for JIDPO in the 21st century continues, as we ready the platforms for design that will take the initiative.

Our Aim

The ultimate goal in promoting design is to complete an environment in which anyone can freely take charge of the thinking and methodology for ingenious solutions to any number of challenges or issues. This action, in fact, represents one definition of designing.

When difficult problems are encountered, all human thinking is funneled into design thinking. Because this shift is quite natural and subconscious, we don't recognize the boundlessness that arises in this thinking. By simply re-validating this thinking in our daily lives as design thinking, and by consciously applying the knowledge thus gained, we can deepen our understanding of the challenge or issue faced, and derives a munificent solution.

The power of design extends further. Through their embodiments (image and form), designs executed by one individual summon designs dormant in the minds of others. The moment a designer is exposed to, and resonates with, an embodiment, he or she is naturally empowered to design his or her subject. In short, a single design can awaken many designs. If accelerating this summoning power held by design were possible, then the resulting synergies could generate a potent force in advancing society as a whole.

While design offers wisdom for ingenious problem solving, design could also become an enormous engine that empowers society as a whole. There in lies a strong reason to promote design.

Our Role

The communities we take part in are filled with vexing challenges and mind-boggling problems, as typified by global environmental issues, and we apparently have no answers. We find ourselves almost ready to throw in the towel, yet historically, humankind has evinced unbelievable strength when faced with seemingly insurmountable problems. Somehow, somebody discovers a brilliant solution, and when the solution is shared by many, humanity overcomes the challenge. This kind of result may not necessarily follow objective methodology. Yet the solution convinces everyone. I believe that the advances of modern society have hollowed out these intrinsic circuits for sharing among people, and have aggravated the severity of our social agenda.

Couldn't the summoning power of design be leveraged, therefore, to restore these intrinsic circuits for sharing? The innate power belonging to design could elevate design into a medium for sharing and solving issues across society, and take design beyond the answers provided in response to each discrete design project. As a medium, design could fulfill a social role that transcends problem solving on an individual basis.

In this context therefore, the role of an institution dedicated to promoting design remains steadfast to the status-upgrade of design into a medium, and the resulting restoration of intrinsic circuits for sharing among people.

First, we delicately draw attention to the issues requiring solutions. We encourage self-directed design activities across society. We then lead the countless designs created to a forum of mutual inspection, and create launch pads for resolving the social issues and "insurmountable" challenges. The resulting reconnections of intrinsic circuitry and improvement in individual design competency reflect the measures that, in fact, bring greater affluence to our lives, and most beneficially and intelligently resolve our outstanding social issues.

Promotion Methodology

More specifically then, how can we improve individual design competency? The question calls for a strategic perspective in terms of determining the central player targeted for this performance upgrade. Today, who needs design most acutely, those in industry, or consumers? The promoting institution first selects the target for promotion, and then implements policy to effectively develop their design competency through design instruction, training seminars, exhibition opportunities, and design competitions, for instance. The programs may vary, but in each case, the promoting institution is engaged together with people (i. e., outstanding designers) who can sufficiently release the summoning power of design.

We should understand designers not only as people gifted in expression of form, but people who can awaken design skills in others and cause them to improve. Designers enlist the cooperation of people facing challenging and tough problems and proceed to design a solution, or a hypothesis. Based on the hypothesized illustration, many other players develop their own designs. And the designer makes progress on his own design, too. Innumerable iterations follow. The mutual appraisal of a spectrum of designs enables the players to acquire more sophisticated design capabilities.

The design advocacy institution's role is not to present a likely solution, but to create design-encountering opportunities for the parties beset with challenges that require solutions. These parties can assist in the design process themselves, in such a way that all sorts of people participate in design. This effort can be described as designing a theater.

The small theater gradually expands to a larger theater, as design competency rises. An important point here is to ensure a successful audience for the theatrical presentation. As the challenge deepens and expands, the solution for the challenge becomes harder to execute solely by the actors in the theater. By creating a structure where on-stage players and audience are united, a design community can evolve, and tough problems can be shared and solved together.

Core Text 11

The Nature of Design

Bionics: Drawn from Nature

Major Jack Steele of the US Air Force coined the term bionics in 1960 to describe what was then an emerging research approach at the interface between natural and synthetic systems. Steele defined bionics as "the analysis of the ways in which living systems actually work and having discovered nature's tricks, embodying them in hardware."

Many other definitions can be found in the literature, but for the purposes of this article I have chosen the following:

Bionics is the derivation of engineering principles employed in natural systems, and the application of these principles to the design or improvement of materials and technological systems.

Biological systems are characterized by their miniaturization, their sensitivity, their high degree of flexibility, their ability to adapt to changing environments, and their high degree of reliability. These design features and the engineering principles that drive them offer a great range of possibilities for research into the improvement of manmade systems.

The concept of mimicking nature very likely dates to prehistoric times. One can envision prehistoric humans fashioning a weapon resembling the claws of the wild animals they fought, or imitating their natural surroundings for the purposes of camouflage.

Leonardo da Vinci, however, may have been tile first true bionics researcher. Many, if not most, of his designs were based on observations in nature. Take, for example, the "ornithopter", a flapping-wing aircraft patterned on his careful anatomic studies of birds (See Figure 4.6). Similarly, the Wright brothers created stabilizers for their airplanes by analyzing how a turkey vulture uses its body to reduce turbulence.

Another example of early contributions derived from the mimicking of nature extends beyond the field of architecture. The South American water lily, Victoria Regia (See Figure 4.7), inspired nineteenth-century architect Sir Joseph Paxton's design of the Crystal Palace in Hyde Park, London.

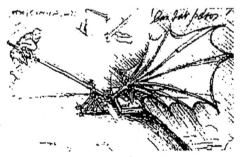

Figure 4.6　Leonardo's ornithopter wing

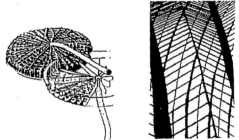

Figure 4.7　Detoil of the ribbed structure of London's Crystal Palace, inspired by the structure of a South American water lily

This plant floats on delicate leaves up to two meters in diameter, yet it is able to support a weight of 90kg. The underside of the leaf carries a system of hollow ribs that give both strength and buoyancy.

As Paxton wrote, "Nature was the engineer, nature has provided the leaf with horizontal and traverse girders and supports that I, borrowing from it, have adopted in this building."

Other examples include the work of Ignazio and Igo Etrich, who built the first tailless glider. Their design followed observations of the propagation of anemophilous plants, whose seeds, transported by the wind, are able to cover considerable distances. Max O. Kramer's anti-turbulence linings for submarine devices imitate the skin structure of dolphins. Even more famously, Georges de Maestral, after being struck by the way in which burdock seeds stick to the coats of animals, conceived the idea for Velcro. NASA and some US military agencies have investigated the biology of nocturnal predators, who are able to attack prey while flying in utmost silence, and the neuromuscular coordination of certain species of beetles capable of memorizing the structure of the terrain they have covered.

Evolution and Design

Through evolution, nature has refined its forms, processes, and systems over a multitude of incremental steps. The complex interplay of evolutionary forces assures the quality of the resulting system.

This lengthy process has resulted in highly adapted systems. Indeed, the principle of evolutionary convergence states that living creatures, even if they are quite different from each other, may have specific structures in common, developed as a result of adaptation. The processes of adaptation to and selection for the respective environments have determined, generation after generation, the development of analogous characteristics. For example, prehistoric ichthyosaurs, the shark, and the dolphin (reptiles, fish, and mammals, respectively) share similar characteristics. Over centuries, evolution has determined that characteristics suitable for an aquatic environment are a streamlined shape, fins in place of limbs, and a caudal fin for stability. Learning from nature, we should reference these structures as ideals to emulate in our designs.

In his book *Chance and Necessity*, biochemist Jacques Monod made a distinction between the two moments of the evolutionary process. The first appearance of certain structures is a chance phenomenon. Subsequently, adaptation, or refinement, is necessary if the structure is to be successful in consecutive generations. One might suggest that this is also the standard process of technological design. Bionics, however, works to reinvent the design process to start with the necessity and look back at nature to find the answer. Thus, you might say, we can skip the chance phase and assume that nature has already tested the existing principles and mechanisms for us.

Because bionics deals with the technical transformation and application of the structures, procedures, and developmental principles of biological systems, it has become an interdisciplinary field that combines biology with engineering, architecture, and mathematics. Over the last decade, with our increased focus on nature, the field of bionics has exploded. Today features derived from nature are commonly used for design improvement—even for marketing strategies.

Bionics can be classified under five main categories:

1. Total mimicry: An object or a material or chemical structure that is indistinguishable from the natural product (for example, early attempts to construct flying machines)

2. Partial mimicry: A modified version of the natural product (for example, artificial wood)

3. Nonbiological analogy: Functional mimicry (for example, modern planes and the use of airfoils)

4. Abstraction: The use of an isolated mechanism (for example, fiber reinforcement of composites)

5. Inspiration: Used as a trigger for creativity (for example, the design for London's Crystal Palace inspired by water lily)

Bio-design: Design Solutions in a Natural State

Bio-design considers the internal and external architecture of "living machines" as extremely efficient design solutions developed to perform multiple role functions in their environments. Probably the earliest type of design methodology used by humans, examples of bio-design are abundant throughout our history. Take the area of transportation design. The use of a fish form for a boat hull or a submarine and the form of a flying bird for the basic configuration of an airplane are not coincidental design solutions. Interestingly, these adaptations were made with very little, if any, scientific knowledge of fluid dynamics; the pioneers who designed them relied only on their conviction that the shapes of these living machines were the forms best suited to perform in their environments.

Nature's highly advanced designs have a common objective: the harmonization of form and function, achieved through the balance of internal and external forces acting on the natural system and the integration of several functions into the form.

An example of design integration is clearly observed in the analysis of the shape of a powerful swimmer such as the shark (See Figure 4.8). The shark's mouth, gills, eyes, and skin sensors are integrated into its form in such a way as to make interference with its movement minimal, yet the function of these organs is unimpeded. Its skin has many functions. It is a heat exchanger, an environmental sensing device, and a self-sealing tank, among others.

The fact that bio-designers refer to the forms or structures of the living world as design models is based on the abundant evidence that shows these forms or structures to be accurate mathematical representations of the balance between environmental and functional forces exerted on the creature, and not the result of random or spontaneous events.

The processes involved in bio-design approach can be divided into four stages:

Figure 4.8 Design integration in action. The shork is a living machine perfectly suited to its environment

1. Select features of a living organism that exceeds current technological capabilities.

2. Derive principles and processes responsible for their superiority.

3. Develop models and methods to describe biological systems in terms useful to designers.

4. Demonstrate the feasibility of translating this knowledge into dependable and efficient hardware.

Biomechanics

Biomechanics describes the use of the laws of physics as well as engineering concepts, to describe the motions of parts of the human body, and the forces acting on these parts, during normal daily activities. The interrelations between force and motion are important, and they must be understand and applied when conceptualizing and designing products.

Anthropometry and Ergonomics

Anthropometry can be defined as the study of human body measurement for use in human related classification and comparison, and is a part of ergonomics.

Designing for people involves making accommodations for their comfort, contentment, freedom, and productivity. The problem of accommodating all people is an ideal goal that may never be reached. Ideally, a design will work for the largest and the smallest male or female adult, but it may also have to be acceptable for use by children and, in many cases, handicapped people.

It is in these types of cases that anthropometric science is helpful. During the course of my career, I have implemented these approaches in the development of products such as sporting tools, cooking systems, furniture, automobile interior, motorcycles, oral care and shaving systems, writing instruments and, most recently at Motorola, wearable communication devices.

Two Bionic Designs

To illustrate the power of bionics, I will explain two designs from my work within the field. The first project came from CAMP, an Italian sports equipment manufacturer, which needed a design for a multifunctional ice axe to be used in variable positions-lightweight, with very high structural strength and a powerful grip structure. The tool would be used under difficult conditions for both mountain climber and materials. In fact, it would have to suffer the extremes of 5,000-meter altitudes and temperatures on the order of $-20℃$. The axe needed to be strong enough to penetrate the ice, but as light as possible so that it did not fatigue the mountain climber.

The natural model I selected was the woodpecker (See Figure 4.9): a bird that chisels into wood to get at the insect larvae on which it feeds. This bird has an extraordinary aptitude for chiseling. It can do 25 hits a second, with a force of impact of $25km/mm^2$.

The woodpecker's body is designed specifically for this movement. Bracing itself with its tail, which functions as a spring, it can take advantage of its center of gravity and the configuration of the bones of its skull to absorb considerable stress. Thanks to this set of characteristics, a woodpecker can utilize its whole body to increase the efficacy of the percussion. These birds do not hammer on the wood by simply moving their necks! Even more incredibly, most woodpeckers weigh only about 500g, or a little more than a pound.

The ice axe I designed (See Figure 4.10 and Figure 4.11) consists of an inner core of titanium, into which is inserted an adjustable aluminum point. These two parts are attached by a hinge inspired by the two valves of a mollusk. Special attention was dedicated to the shape of the handle. Rather than designing it to be straight, I incorporated into it a slight curve, again taking the body of the woodpecker as a model. This improves the efficiency of the blow.

Figure 4.9 The strength, balance and light weight of the woodpecker inspired o design for on ice axe.

Figure 4.10 and Figure 4.11 This ice axe consists of an inner core of titanium, into which is inserted on adjustable aluminum point. These two parts are attached by a hinge inspired by the two valves of a mollusk. Special attention was dedicated to the shape of the handle, whose form is curved rather thon straight, as the woodpecker's body is curved rattler thon straight. This improver the efficiency of the blow. The tool is also lightweight, to conserve the energy of the user.

The handle is lined with a knurled layer of PBT rigid polyester and is covered with an elastic layer of Rynite to provide the grip. For this component, I took inspiration from the epidermis of sharks: rigid elements overlying a soft base. The overall result is a structure that withstands heat stress, water, damp, and UV radiation.

This design strategy led CAMP to change its image, its line of products, and its marketing strategy. Emphasizing the environmental aspect of the design (See Figure 4.12) grabbed customers' interest and actually had an impact on the whole industry sector.

The second project I want to mention was a new line of rugged, shock-resistant, weather resistant phones introduced this past December by the Motorola division that designs technology and handsets for Nextel Communications. My model here was the tough protective exoskeleton of lobsters and other crustaceans.

The outer shells of these animals are constructed of hard and soft layers of chitin combined with calcium carbonate. The layering provides a covering that protects internal organs extremely well. To achieve the same effect and to protect the phone's inner workings, our design group used hard and soft layers of polymers (chemical compounds with long repeating chains of atoms) to cover the entire exterior of the phone (See Figure 4.13). The layers are made with substances such as polycarbonate and Santoprene, a rubberlike plastic material.

Figure 4.12 The environmental design of the wood pecker was intriguing to customers.

Figure 4.13 This series of rugged phones developed for Nextel incorporated biodesign principles derived from the tough protective shells of lobsters and other crustaceans.

Key Words

[1] bionics [baɪˈɒnɪks] n. [美] 仿生学

［2］synthetic [sɪn'θetɪk] adj. 1. 合成的，人造的 2. 假的，非天然的；虚伪的 n. 合成物，合成纤维，合成剂

［3］buoyancy ['bɔɪənsɪ] n. 1.（物体在液体里的）浮性 2. 浮力

［4］evolution [ˌiːvə'luːʃən] n. 演变；进化；发展

［5］emulate ['emjʊleɪt] vt. 1. 与……竞争，努力赶上 2. 计算机程序等仿真；模仿

［6］mimicry ['mɪmɪkrɪ] n. 模仿；模仿的技巧

［7］harmonization [ˌhɑːmənaɪ'zeɪʃən] n. 调和化，一致，融洽

［8］unimpeded [ˌʌnɪm'piːdɪd] adj. 无障碍的；无阻挡的

［9］organism ['ɔːgənɪzəm] n. 1. 有机物，有机体 2. 有机体系

［10］biomechanics [ˌbaɪəʊmɪ'kænɪks] n. 生物力学

［11］anthropometry [ˌænθrə'pɒmɪtrɪ] n. 人体测量学

Key Sentences

1. Bionics is the derivation of engineering principles employed in natural systems, and the application of these principles to the design or improvement of materials and technological systems.

仿生学起源于自然系统中的工程原则，并将这些原则运用到设计中或改进材料和科技系统。

2. Biomechanics describes the use of the laws of physics as well as engineering concepts, to describe the motions of parts of the human body, and the forces acting on these parts, during normal daily activities. The interrelations between force and motion are important, and they must be understand and applied when conceptualizing and designing products.

生物力学描述物理规律的使用，以及工程的概念，还描述了在正常的日常活动中人体各部位的运动和力。力和运动之间的相互关系是重要的，当构思和设计产品时必须理解和应用这一点。

课文翻译

设计的本质

仿生学：师法自然

美国空军少校杰克·斯蒂尔在1960年创造了仿生学这个词，来形容当时自然与人造系统界面之间的一种新兴研究方法。斯蒂尔将仿生学定义为"仿生学是研究生物系统实际工作方式、发现自然的技巧、并且在硬件中表现它们"。

在文献中还可以找到许多其他定义，由于本文的目的，我选择定义如下：仿生学是自然系统中工程原则的起源，并将这些原则运用到设计中或改善材料和科技系统。

生物系统的特点是微小型化、敏感性、高度的灵活性。它们能够适应变化的环境，具有高度的可靠性。这些设计特征和工程原理促使它们为人造系统改善的研究提供很大的可能。

模仿自然的概念可以追溯到史前时期。人能想象出史前制定的武器和他们与之战争的

野生动物的爪子很类似，或是为了伪装的目的模仿自然环境。

列奥纳多·达·芬奇是第一个真正的仿生学研究者。他的设计很多是基于观察自然的基础上的。例如，"扑翼式飞机"是效仿他对飞鸟细致解剖的研究。同样，怀特兄弟通过分析红头美洲鹫如何利用它们的身体保持平衡为他们的飞机创造了稳定器。

还有超越建筑领域的起源于模仿自然的其他案例。维多利亚睡莲激发了19世纪建筑师帕克斯顿伦敦海德公园的水晶宫设计。

这种植物的叶子漂浮在直径两米的范围，但还能支持90公斤的重量。叶子下面一系列中空的肋骨，能够提供强度和浮力。

如帕克斯顿所写，"自然是工程师，自然提供给叶子水平的和遍布的梁来支撑它们，借鉴于它，适用于本建筑。"

其他例子包括伊尼亚齐奥和伊戈·艾垂奇的作品，他制作了第一个无尾滑翔机。其设计基于对风媒（传粉）植物繁殖的观察，其种子由风传播，能够覆盖相当大的距离。马克斯·克莱默阿的抗波动潜艇设备衬里模仿海豚的皮肤结构。更著名的是乔治·德·麦斯特拉尔在受到牛蒡种子粘贴于动物皮毛这种方式的触动，构思了维可牢尼龙搭扣的想法。美国航天局和美国的一些军事机构调查了夜间捕食者的生物学，他们能在极其安静的飞行中攻击猎物，这和某种能够记忆其所覆盖地形结构的甲虫神经肌肉相协调。

进化与设计

通过进化，大自然经过一系列渐进的步骤已经完善了其形式、程序和系统。进化力量复杂的相互作用保证了最终系统的质量。

这个漫长的过程导致高度适应的系统。事实上，进化趋同的原理阐述了居住的生物，即使它们彼此区别很大，也可能具有特别的结构共性，发展为具有适应性的结果。对各自的环境适应和选择过程已经确定，经过一代又一代，发展为类似的特征。例如，史前鱼龙、鲨鱼和海豚（分别为爬行动物、鱼类和哺乳动物）有着相似的特点。几个世纪以来，进化决定了其特征为适合水环境的流线形态，四肢为鳍，以及一个稳定的尾鳍。师法自然，我们以这些结构作为典范在设计中仿效。

在他的著作《偶然与必然》中，生物化学家贾克·莫诺将进化过程的两个时刻加以区别。某些结构的首次出现是一个偶然现象。如果结构要在连续几代获得成功，随后的适应或细化是必要的。有人可能会说，这也是技术设计的标准过程。然而，仿生学重新发明设计过程，从必要性入手，回顾自然，寻找答案。因此，你可能会说，我们可以跳过机会阶段，并假设自然已为我们测试现有的原则和机制。

由于仿生学涉及技术改造和结构、程序的应用以及生物系统发展的原则，它已成为一个跨学科领域，它将生物学与工程、建筑和数学相结合。在过去的十年中，我们更加关注自然，仿生学领域已经被突破了。今天，来自大自然的特征被普遍用于设计改进，甚至用于市场策略。

仿生可分为以下五大类：

1. 共有模仿：一个物体或材料或化学结构与天然产物难以区分（例如，早期试图建构飞行器）
2. 局部模仿：对自然产品的一种修改版本（例如，人造木材）
3. 非生物的比喻：功能模拟（例如，现代飞机与机翼的使用）
4. 抽象：一个孤立的机制的使用（例如，纤维增强复合材料）

5. 启示：用作触发创意（例如，伦敦的水晶宫设计由睡莲激发灵感）

仿生设计：自然状态的设计解决方案

仿生设计认为，"生物机器"的内部和外部结构作为极为有效的设计开发解决方案，在其环境中执行多个角色的功能。它可能是人类使用的最早的设计方法学类型，贯穿历史，生物设计实例丰富。以交通设计领域为例。鱼的形式用于船体或潜艇和飞鸟的形式用于飞机的基本构造并不是巧合的设计解决方案。有趣的是，这些修改非常少，如果有的话，也是流体动力学的科学知识，设计它们的先驱者只能依靠自己的信念，这些活的机器形态最适合其生存环境。

大自然的高级设计有一个共同目标：形式与功能的统一。通过自然系统中内部和外部力的平衡和功能与形式的整合来完成。

有一个设计整合的例子可以分析一个强大游泳者的形态，如鲨鱼被清楚地观察时。鲨鱼的嘴、鳃、眼睛和皮肤传感器被集成于它的形体上，通过这种方式可以使它的运动干扰很小，但这些器官的功能是不受阻碍的。它的皮肤有很多功能，是一个换热器、环境检测设备、自密封罐等。

事实上，仿生设计者参考生物世界的形式或结构作为设计模型是基于大量的证据基础上的。这些证据表明这些形式或结构是影响生物的环境与功能的准确数学表述，而不是随机的或自发事件的结果。

仿生设计方法所涉及的过程可以分为以下四个阶段：
1. 选择超过目前技术能力的生命有机体的特征。
2. 为其优越性负责的派生原则和过程。
3. 发展模式和方法以对设计师有用的术语来描述生物系统。
4. 表明将这种知识翻译成可靠和有效率的硬件的可实施性。

生物力学

生物力学描述物理规律的使用以及工程的概念，描述了在正常日常活动中人体各部位的运动和力。力和运动之间的相互关系是重要的，当构思和设计产品时，它们必须被理解和应用。

人体测量学和人体工程学

人体测量学可被定义为人体测量研究，这种测量研究用于人类相关的分类和比较研究，是人体工程学的一部分。

为人而设计包括为他们提供住宿使其舒适、满足、自由和生产率。容纳所有人的问题是一个理想目标，可能永远不会实现。在理想情况下，设计将为最大和最小的成年男性或女性而工作，但它也可能被儿童以及伤残人士使用。

在这些案例类型中，人体测量学是有帮助的。在我的职业生涯中，已将这些方法实施于产品开发，例如，运动工具、烹饪系统、家具、汽车内饰、摩托车、口腔护理和剃须系统、书写工具以及最近在摩托罗拉公司的可穿戴通信设备。

两个仿生设计

为了说明仿生学的作用，我将在此领域解释我的作品中的两个设计案例。第一个项目来自意大利体育器材制造商CAMP，它需要一个多功能的冰镐设计，用于变动的位置，质量轻，具有很高的结构强度和强大的抓牢结构。该工具将用于登山者和材料处于困难条件下。事实上，它必须承受5千米的海拔高度和零下20℃的温度。斧头需要强大到足以穿透冰层，但尽可能足够轻，可使登山者不感觉疲劳。

我选择的自然模型是啄木鸟。这种鸟能够凿入木材，获得寄养其中的昆虫幼虫。这种鸟具有非凡的凿穿才能。可以每秒钟敲击 25 下，冲击力为 25 千米/平方毫米。

啄木鸟的身体专为这一运动而设计。用尾部支撑自己，它的功能相当于一个弹簧，它可以利用其重心优势和脑部骨骼配置以吸收相当大的压力。这一组特征下，一个啄木鸟可以利用它的整个身体提高打击的效率。这些鸟儿不是通过简单的移动脖子敲击木头！更令人难以置信的是，大多数啄木鸟质量只有约 500 克，或略超过 1 磅。

我设计的冰镐包括钛内核，在其中插入一个可调铝点。这两部分是由一个铰链连接，它是由软体动物的两个瓣膜受到启发。特别注意的是手柄的形状。不是将它设计成直的，我加入一个微小的曲线，再一次以啄木鸟的身体为模型。这提高了打击效率。

把手内衬 PBT 刚性聚酯纤维滚花层，并覆盖了有弹性的热塑性聚酯树脂提供抓握力。对于这一部分，我从鲨鱼的表皮获得灵感：刚性元素覆盖着柔软的表层。总的结果是一个抵抗热应力、水、潮湿、紫外线辐射的结构。这种设计策略导致 CAMP 改变自己的形象、产品线和市场营销策略。强调设计获取客户兴趣的环境因素，实际上已经对整个行业产生影响。

第二个项目我要提到一个坚固、抗震、抵抗气候的手机系列。由摩托罗拉分公司在去年 12 月引入，为纳克斯泰尔网络通信公司设计技术和手机。这是我的模型，它具有龙虾和其他甲壳类动物强硬的保护骨骼。

这些动物的外壳由坚硬和柔软的角质层混合碳酸钙组成。分层提供了一个覆盖层，相当好地保护了内部器官。为了达到相同的效果，并保护手机的内部运作，我们的设计组采用聚合物软，硬层（带有原子重复链的复合层），覆盖电话的整个外部。该层是由诸如聚碳酸酯和山都平等材料组成，山都平是一种类似于橡胶的塑料材质。

Free Reading 1

Design for Need

With ever-increasing levels of consumption and disposable income in the decades following the end of the Second World War, there were a growing number of critical voices including Vance Packard, Victor Papanek, and Richard Buckminster Fuller. Such critiques of industrialization and consumption had been long-standing, from Victorian writers such as John Ruskin in his critique of material over-indulgence, *The Stones of Venice*, continuing through the 20th century to the writings of Naomi Klein, particularly *No Logo: Taking Aim at the Brand Bullies* (2000). The design profession was understandably slow to respond to many of these concerns, implicated as it was in mediating between producer and consumer.

In 1969 ICSID (International Council of Societies of Industrial Design) held a conference in London entitled "Design, Society and the Future" at which leading designers reflected upon the social, moral, and economic consequences of their actions.

The oil crisis of 1973 brought about by the Middle Eastern War, coupled with the inability of American technological superiority to bring the Vietnam War to a speedy resolution raised a number of fundamental questions about the nature of progress. Once again the design profession sought to address such issues through another ICSID conference. Mounted in London and entitled "Design for Need", the Third World, alternative technology, and design for disability were among the topics addressed. Design for disability was not widely practiced though one major Swedish consultancy, Ergonomidesign, specialized in many aspects of the field. Two of its leading members, Maria Benktzon and Sven-Eric Juhlin, were pioneers in the field and sought to devise aesthetically pleasing and stylish design solutions for everyday products, so bringing the disabled into the mainstream of consumption. (By 2003 Ergonomidesign employed 27 industrial designers, engineers, and ergonomists.) From the 1970s onwards debates about design for need were increasingly concerned with environmental and ecological questions, stimulated by growing concerns about the finite nature of fossil fuels and the consequences of global warming.

Free Reading 2

The Future of Design

Sustainable Design (Green Design)

The key principle for green world is, in a word, sustainability. The well-being of today's generations is not increased at the expense of future generations... Each generation should ensure it passes on to the next a stock of assets no less than that which it inherited.

Designers therefore have a crucial role to play in achieving a more sustainable economic and social order. The complexity and importance of the designer's role is highlighted. Designers must ensure that by providing solutions to one set of environmental problems they are not creating or increasing others.

The definition of green design

- ◆ Green design is a term describing the various techniques used in prioritizing environmental considerations through the various design stages of a product or system, with the objective of conserving or minimizing any damage to the environment.
- ◆ Green design is a design philosophy that is increasingly becoming a mainstream practice. That our built environment has a profound impact on our natural environment, economy, health, and productivity makes this even more imperative.
- ◆ All products have some environmental impact, though some products use more resources, cause more pollution or generate more waste than others. Consequently the objective of green design is to identify those, which cause least damage.

The principles of green design

The fundamental principles of green design focus on using products with recyclable materials and recycled content, utilizing least toxic materials and manufacturing processes, minimizing or designing away the extraneous and designing for durability and longevity.

The key components of green design are summarized below:
- ◆ Integrate design aspects for multiplicity of function;
- ◆ Design for durability and longevity—think about the unintended consequences of maintenance and renewal;
- ◆ Select materials that use their base resource most efficiently;
- ◆ Value long-term benefits over short-term profits;
- ◆ Use products with recyclable materials and recycled content;
- ◆ Reduce, reuse, recover and recycle;
- ◆ Design to use only local and regional resources;
- ◆ Look for least toxic materials and manufacturing processes;
- ◆ Take full account of the effects of the end disposal of the product;
- ◆ Minimize nuisances such as noise or smell;
- ◆ Analyze and minimize potential safety hazards.

The benefits of green design
- ◆ Resource efficiency by designing products that use less energy and raw materials in production and consume less energy and resources in use.
- ◆ Reduced cost from using energy and materials more efficiently.
- ◆ Increased profit from more efficient products, niche consumer marketing, and extending products' lifespan.
- ◆ Improved whole systems function—such products have a more efficient manufacturing process, last longer and work better for end users.
- ◆ Increased cooperation among designers, suppliers, and manufacturers can lead to new innovations and better products.
- ◆ Shorter production time due to increased efficiencies.

The reality problems towards the green design
- ◆ Initial costs arise from investment in time, materials, new equipment, and other items.
- ◆ Resistance to change—manufacturers, suppliers, and even consumer's may resist changing a product design, having to work with new materials, doing things a new way, or seeing their product look a different way.
- ◆ Consumer indifference—if the product is an unknown brand, or significantly more expensive than a non-green product, consumers may opt not to buy.

The key players in green design
- ◆ Designers

It is designer's role to set the tone for product features and functions based on the green concept.
- ◆ Universities and professional design associations

It offers cutting-edge green design training programs.
- ◆ Manufacturers
  - • They contribute to the design process by determining what is producible, de-

termining efficient manufacturing processes, and by working with designers on other product elements.
- They must buy into the green design concept and commit to working within the manufacturing constraints on green-designed products.

◆ Consumers
- Consumers create demand for green products.
- Many consumers must be educated about environmental attributes through product labeling and marketing, while others have let companies know they will only spend money on products that have been designed for the environment.

Action plan
◆ Fill a consumer's need. While many consumers might buy a product for its green attributes, most will not buy a product that does not meet a specific need.
◆ Determine what that need is and how to design a green product to fill that need.
◆ Establish a design team of design team of designers, marketers, manufacturers, suppliers, and others along the production chain to develop design ideas.
◆ Evaluate the options by using life-cycle analyses, cleaner technologies, substitute assessment, and other tools.
◆ Determine production costs and precisely which product components should be designed with environmental attributes.
◆ Market the product and work to get it certified as environmentally friendly by ecological agencies such as green seal. Many consumers choose products that meet environmental standards and like proof that a product meets those standards.

Towards the green version
◆ Just as good design is the cornerstone of a successful product, green design is the cornerstone of a good, green product.
◆ Emerging trends suggest that green is the way to go for a nation in the long term development.
◆ Disposition by designing it well to begin with, producers and manufacturers can step ahead of legislation requiring manufacturer responsibility for the disposition of a product.

Future Design
The reasons for taking a version of future in design
◆ Advanced societies are facing saturated markets, increasing prince erosions, and a situation where no single developer has sole "ownership" of any new technology. In emerging markets, however, the situation is more fluid, making it difficult to decide which direction to take. That's why it is vital to spot market trends as early as possible.
◆ With the version of future you can anticipate what the future holds.

- You need to be able to recognize the direction people to moving in, and identify—and even trigger—new inspirations and needs.
- Vision of the future explored and visualized possible lifestyle scenarios in the domestic, mobile, personal and public domains. Five years later, almost two-third of all the product concepts featured in the product has become reality.

Human centered design

Taking the notion of human centered design is another version for future design. The definition of Human Centered Design is as follows:

- Human Centered Design (HCD), or User Centered Design (UCD), is all about designing products tailored to users.
- A product designed with HCD should hide the technique and details of how it works and be completely intuitive to use.
- Human Centered Design is a methodology that maximizes the likelihood that a product will meet a user's wants and needs, behave the way they expect it to, and provide them with a quality user experience that will not only satisfy, but also delight them.

The reasons for human centered design

- Today it is not enough just to build into a product implicit quality, such as excellent functionality and usability, and explicit qualities, such as an appearing look and feel. You need to discover new, attractive qualities by considering the whole experience of interaction between user and product over time.
- Making the "connection" between people and objects is essence of design in the New Economy.
- Technology is pervasive. The gadgetry of the 1980's is behind us. The explosion of network in the 1990's has delivered the Internet. The Internet has changed our lives.
- Humanity is the new frontier. Simplicity and usability should be embraced as core design values. Emotion and intuition should drive the process.
- It assumes that all the participants in the design process bring their own personal bias into the process and that the actual end-users are really the only participants who can even come close to providing objective input as, after all is said and done, they are ones that are going to use the product.
- Human centered design considers all the factors—physical, cognitive, social, culture, and environment—that influence people's interaction with their world. The interaction between people and environment are shown in Figure 4.14.

Figure 4.14 The interaction between people and environment

The value of human centered design
- According to a study conducted in 1993, it can cost 10 times more to correct user interface problems in development than in design and 100 times after the product is released to market. It's often a lot more expensive, if not impossible; to make changes after a product is completed.
- Proper application of Human Centered Design methodology will be completely invisible in the final design solution but its absence won't be.
- Usable products are desirable products and this obviously makes good business sense.
- A good quality user experience=Increased business value
- A good quality user experience increase sales and customer satisfaction while enhancing product and company brands.

The principles of human centered design
The overall principles of Human Centered Design include:
- Understand the problem:
  - Determining the target market, intended users, and primary competition is central to all design and user participation.
  - You must know whom you're designing for.
- Understand users:
  - A commitment to understand and involve the intended user is essential to the design process.
  - If you want a user to understand your product, you must first understand the user.
- Assess the competition:
  - Successful design requires ongoing awareness of the competition and its customers.
  - Test your user's tasks against the competition. You cannot design successfully in isolation.
- Design the total user experience:
  - Everything a user sees and touches should be designed in concert.
- Evaluate designs:
  - User feedback is gathered early and often, using prototypes of widely ranging fidelity, and this feedback drives product design and development.
- Manage by continual user observation:
  - Throughout the life of the product, continue to monitor and listen to your users.
  - Let your user's feedback inform your responses to market changes and competitive activity.

Free Reading 3

Design on the Chair

Over the last 150 years, the evolution of the chair has paralleled developments in the architecture and technology and reflected the changing needs and concern of society to such an extent that it can be seen to encapsulate the history of design. As George Nelson pointed out in 1953, "every truly original idea—every innovation in design, every new application of materials, every technical invention for furniture—seems to find its most important expression in a chair." In our times, this is nowhere more apparent than in the development of better performing, more ergonomically refined chairs. The highly competitive office seating market, in particular, demands continual technical advances, as it is increasingly driven by tougher health and safety legislation and corporate specifiers ever more mindful of the welfare of their workforced.

Achieving a good solution to the problems posed by the chair is a complex and challenging proposition, even though, over its long history, its function as an aid to sitting has remained virtually unchanged. Chairs support people of all different shapes and sizes for different lengths of time and for different purposes, whether it be eating, reading, resting, waiting, writing or office tasking. Furthermore, each sitting position is invested with its own degree of social significance and set of conventions, including rothopaedic constraints. In most cases, the chair must adequately support the weight of the sitter at such a height that the legs hang down and the feet touch the floor. In this conventional sitting position, the weight of the head and torso is carried down to the bones of pelvis and hip, the timeless problem associated with this physical relationship is that however much a chair seat may be softened, the pressure of the bone will eventually be felt on the flesh of the buttocks as uncomfortable. Ultimately, this result in the user having to change position—something that is done on average every ten to fifteen minutes. Indeed, the more exactly a chair is formed to give "ideal" static support and posture to the average human frame, the more it guarantees discomfort and, thereby, psychological stress for people with non-standard anatomies or those who do not wish to assume that particular posture. It is probably safe to say, therefore, that while the facility for correct lumbar support is important, especially in office seating, it is not as crucial as the chair allowing the user to move their legs freely and to make frequent adjustments of posture. For more healthful sitting, a chair should thus facilitate freedom of movement and encourage a variety of postures while providing flexible continuous support.

Beyond the technical considerations of sitting and how well users can physically and psychologically connect with specific forms according to different functional contexts, chairs are also designed an acquired for reasons to do with symbolic content, aesthetics and fashion. Of all furniture types, chairs especially serve to bolster egos and demonstrate "taste", while revealing their owner's sociopolitical viewpoint and real or should-be social and economic status. To this end comfort, practicality and economy have often been sacrificed in favour of representation of decorative styles, radical design agendas and/or the self-expressive impulses of designers.

The Extraordinary diversity of chairs created since the mid-19th century has largely been due to the fact that, owing to the variety of the intended functions of the chair and the anatomic variability of users, there are no ideal forms. There can be many excellent solutions at any one time to the different contexts of use. While the profusion of designs for a specific function may share numerous similarities, at the outset what fundamentally differentiates one from another is the extent to which the designer has viewed function as either the purpose and goal or the subject of the chair. Whether the preference in approach has been weighted towards utility or aesthetics, the primary object of chair design remains the same—making connections—and over the last 150 years there have been innumerable interpretations of hoe best to achieve this. More often than not, the creation of a meaningful solution involves a process which not only takes into consideration intended function, appropriate structure (including deployment of materials) and aesthetics, but also method of manufacture, nature of the market, ultimate cast and proposed appeal. Different chairs emphasise different combination of connections according to the priorities of their designers and the needs and concerns that are being addressed at different times.

As the preoccupations of society change, so too do designers' and manufactures' responses to them. What may be viewed as a rational solution in one period, therefore, may be viewed as exactly the opposite in another. While some designers strive for and achieve an authority which leads to varying degrees of longevity, even those deemed "classic" have a limited functional and aesthetic appeal. Just as tastes change, so too other factors, such as expectations of comfort, vary from period to period and between different cultures. The inherent ephemerality of design, therefore, also accounts for the myriad solutions to the different functional contexts of the chair.

Although there is never one right answer to any given type, some chairs have had an enormous impact on the course of furniture design, for example Marcel Breuer's B3 club chair ("Wassily") of 1925. Alvar Aalto's "Paimio" No. 41 chair of 1931—1932 (See Figure 4.15), Charles and Ray Eames' moulded plywood chairs of 1945—1946 (Figure 4.16), and Joe Colombo's 4860 chair of 1965. See Figure 4.17, Additional Living System, 1967—1968 Textile-covered moulded polyurethane foam on tubular iron frame with metal clamps. These highly innovative designs were born out of

Figure 4.15 Alvar Aalto Paimio. model No. 41, 1930—1931 Bent laminated and solid birch frame with lacquered bent plywood seat section

the search for better, more effective connections-a search which, more than anything else, has progressed design theory and brought a succession of important advances in technical processes and materials applications, from tubular metal to moulded plywood to injection-moulded thermoplastics. Theoretical and technological progress has, historically, not only invigorated interest in chair design but also fuelled the diversity of alternative solution.

Figure 4.16 Charles & Ray Eames Model No. 670, 1956 Rosewood-faced moulded plywood seat shells with leather-covered cushions, cast aluminium base

Figure 4.17 Joe Colombo Additional Living System, 1967—1968 Textile-covered moulded polyurethane foam on tubular iron frame with metal clamps

Architects have always been closely associated with chair design through their abilities to solve problems of structure and to make and exploit connections. In the quest for greater unity of design, architects such as Charles Rennie Mackintosh(1868—1959), Alvar Aalto (1898—1976) and Carlo Mollino (1905—1973) included chairs within their artistic schemes for interiors and buildings. But as the manufacture of chairs moved away from the domain of the craftsman towards that of the industrial process, architects were also ideally positioned, with their background knowledge of engineering, to pioneer innovative chair designs within the constraints of modern manufacturing technology. Chair design has especially appealed to architects, for through it more easily than with architecture; they have been able to communicate their design philosophies in three dimensions. According to the British architect Peter Smithson, writing in 1986, "It could be said that when we design a chair, we make a society and city in miniature. Certainly this has never been more true than in this century. One has a perfectly clear notion of the sort of city, and the sort of society envisaged by Mies van der Rohe, even though he has never said much about it. It is not an exaggeration to say that the Miesian city is implicit in the 'Mies' chair." As a potentially mass-produced and thereby more accessible microcosm of the ideological aspirations of the architect, the chair has allowed some architects to make connections with far more people than would ever use or even view their builings.

Innovation in Practice: The Calor Aquaspeed Iron

By Powell, Dick

In this case study, Dick Powell (See Figure 4.18) translates innovation theory into reality. The challenge was to bring new vision to an old problem—redesigning a steam iron. Researchers mapped design, product, and social trends. They reconfirmed the corporate brand message as successfully engaging customers. They distilled breakthrough opportunities, and then the development team and the CEO made it happen with a design that is changing expectations and exceeding sales projections.

Figure 4.18　Dick Powell, Director

There's a lot of baloney spoken in the name of innovation theory... and some useful truths. After 20 years of designing consumer products, I can offer these observations:

* Innovation requires a consummately well-articulated vision of what you are trying to achieve—one that all parties, from the engineers to the eventual consumer, can believe in.

* Innovation requires at least one person who fully understands all the ramifications of that vision and is armed with the authority and means to make it happen.

* Innovation requires experience, and the insight that comes with it, to balance what's important and what isn't, at any one point in the process.

* Innovation is often not a big idea that changes everything, but rather a series of smaller ideas that fit together in a unique way to create something new and better.

And two more:

* All people are creative, but some people are more creative than others.

* It's never as easy as you thought.

The Calor Aquaspeed iron project reflected a number of these observations.

The Nature of the Beast

Seymourpowell has designed a huge number of products, from trains and cars to digital watches and laptops. When people ask me, "What's the hardest product to design?" I say it has to be the steam iron. Why? Because the complexity of a steam iron's internal workings is wholly inseparable from its external form, which in turn is determined by its ergonomy on the one hand, and its functionality on the other. Factor in the need for ease of manufacture, low cost, and reasonable investment, not to mention on-shelf differentiation in a market in which products are more alike than different as a direct consequence of all this, and you can begin to understand why the word styling (and by that I mean external design) is a wholly inappropriate word. It's a huge design problem ... and that's before you even contemplate being innovative.

Calor is a French company and brand that belongs, along with SEB, Tefal, Rowenta, Moulinex, Arno, and Krups, to the SEB Group. SEB stands for La Socit Emboutissage de Boulogne, which originally started life as a metalworking company making pressure cookers. Having variously bought and acquired Calor and Tefal, SEB became the SEB Group, which continues to sell products under all three of these well-established brands in France, but sells the output of all three companies elsewhere under the name of Tefal (except in the US, where the brand name is the slightly amended and hyphenated T-Fal). As a group, they are one of the world's largest and most successful manufacturers of domestic products and appliances.

Seymourpowell's relationship with the companies of the Group goes back to 1985, when we designed the world's first cordless kettle, the Freeline, for Tefal. In the ensuing 19 years, we have worked with the Group to create a considerable number of innovative products, including deep-fat fryers, vacuum cleaners, beautycare items, irons, kettles, and toasters.

Avantis

The redesign of Calor's Avantis steam iron was Seymourpowell's first linen-care project for the company. We had worked with Calor before, so we already knew that the firm well understood the importance of expressing its brand values and communicating functionality clearly and persuasively through the design of its products. We hoped to continue that tradition with the Avantis project.

Calor irons already had a unique differentiation point: Their soleplates, which in other irons are typically made of stainless steel or anodized aluminum, are enameled. This gives them a finish that makes them very durable and resistant to scratching and discoloration. Even more important, however, is the feet that printing two different colors of enamel—one as a base colour and the second overprinted as a series of lines and details—allows the iron to float on the ridges created by the two-color surface, decreasing friction and increasing its "glide" and hence the speed of ironing. Calor designers made the most of this by aiming for a streamlined, dynamic, "fast" look.

For Avantis, Seymourpowell modernized the fast look by removing its edginess and softening it. We created a highly streamlined, dynamic, contemporary iron with a new soleplate design that featured a point at the back of the iron to help part the fabric on the backward stroke (See Figure 4.19).

Steam Generators

Seymourpowell then moved on to other categories of iron and linen-care products, including a complete rethink of the steam generator, a category that Calor had pretty much established. (For those who have not seen or used a steam generator, they improve and speed the ironing process because they produce much more steam.) Previous generators

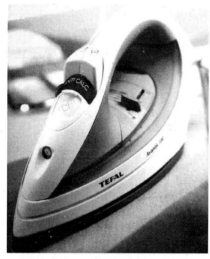

Figure 4.19 Avantis, Seymourpowell's first iron design for Calor, modernized the "fast" look

were basically boxes with irons plonked on top; the box acted like a glorified kettle. This design creates a fantastically effective ironing system, but there are drawbacks: The iron is unstable, and using it is tiring, since the user has to lift the iron back up onto the steam-generator box at the end of each burst of usage.

To combat these problems, Seymourpowell reconfigured the generator's internal architecture to achieve an inclined iron "rest" at a significantly lower level (See Figure 4.20). This proved to be ergonomically better (less lifting and wrist rotation), and it also allowed the iron to be nestled into the "box" for greater stability and safety. The product, available in two variants called Calor Express and Calor Pro Express, was more integrated, both visually and in use, than its predecessors and the competition. It changed the whole game in the category, increasing sales and market share, and in 2002 won for the Seymourpowell/Calor team a DBA Design Effectiveness Award.

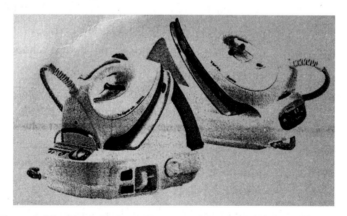

Figure 4.20 The Color Pro - Express and Express generators (front and rear, respectively) won a DBA (Design Business Association) Design Effectiveness Award for Calor and Seymourpowell

Avantis 2

After the design of further ranges of both high-end and low-end irons, the moment finally came, four years after its creation, to reconsider Avantis-the mid-line range. It was a moment to pause and reflect on where we were and where we were going. As a business, Calor is passionate about innovation. It spends time and money looking for ideas to give its products a competitive edge. The shelves in the company's R&D department testify to that. They groan under the weight of prototypes and mock-ups of every conceivable idea.

It's not for lack of trying that very few make it; it's because it's so difficult to produce a new and unusual product but still sell it at a competitive cost (especially alongside Chinese products). As is so often true, the process of developing new products is analogous to pushing a big boulder over rocky terrain-much time is spent in studying which way looks the easiest and most promising, when what's really needed is a helicopter to take an overview and scope out the big picture, and all the factors that might bear upon it—a strategic review.

Enter our research, branding, and strategy team (SPF-Seymourpowell Foresight), which had not been involved in previous Calor design projects. Bearing in mind that the core value of fast had been evolving without any real strategic purpose for close to 10 years, and that the competitive context had changed significantly over that time, the key brand question was-does fast still work as a core value? To map this context and answer that question, I briefed the Foresight team to look at a large number of relevant landscapes for the client, including:

* A timeline showing a visual history of recent developments in the Tefal/Calor range;

* National market analyses of key irons mapped against price and visual sophistication;

* An analysis of the success or failure of brand language used by Tefal/Calor and its competitors;

* A presentation of the opinions of an independent, external panel of design experts;

* An analysis of the most pertinent general, social and product trends;

* A strategic formulation of the way forward, based on this initial analytical and trend work.

Some Answers

The fundamental conclusion was indeed that fast still worked as a basic identity for Calor/Tefal, communicating efficiency and contemporaneity. We believed very strongly as a result of our analysis that fast had established for the brand a relative degree of distinction in terms of brand language within a highly homogenized marketplace, where products are more similar than they are different. Pointing the way forward, we stated:

.... *We need a "beautiful" product which engages emotionally. Beauty is not a word designers are comfortable with, but a beautiful product needs to be harmonious, balanced, proportioned, and express a simplicity of use and purpose.*

The strategic review also identified several potential strands for development that mirrored changing patterns of use among consumers (these must, for the moment, remain confidential, since they are still under development).

In parallel with SPF's research (and that's important—because serial idea generation is never as effective as doing it in parallel, in my view), the creative team began to generate concepts around specific ironing problems that might then be applied to Avantis 2. We hardly needed a focus group to tell us that number one on consumers' wish list was a truly cordless iron. In fact, cordless irons do exist, but current technology leaves them quite compromised in performance, and for cost reasons alone they fell outside our brief. I mention them here only because they illustrate what's wrong with the kind of innovation theory predicated on consumer "insights", which too often allows for the complete dislocation of new ideas from the harsh realities of both science and commerce. Instead, what's truly needed is balance—balance between what you want to do and what you can do. Balance requires extensive knowledge of all the issues surrounding the problem, along with an instinctive feel for how their resolution might physically be realized—long before you can actually try to do so.

Many ideas were generated, some of which have gone into longer-term development at Calor and so can't be discussed here. But again, we didn't need a focus group to identify two immediate problem areas: filling an iron with water through a hole the size of a postage stamp, and stability—to combat the iron's annoying tendency to topple off the ironing board. The solutions too are blindingly obvious: as large a filling hole as possible, and a huge heel for stability. But how to achieve these without being ungainly, ugly, heavy and unwieldy?

Usually, innovation lies neither in identifying a problem, nor in proposing abstract solutions, but rather in embodying those ideas and solutions effectively. This, I think, is the designer's greatest strength—the ability to conceive a vision of the whole quickly and fluidly, without losing sight of the myriad factors that ultimately affect its resolution.

In contrast, my experience of working with talented creative engineers is that they are rarely at their best trying to resolve an abstract problem. But give them the same challenge armed with a credible and compelling vision, with which they can understand exactly what is needed and in what form, and they will examine an array of potential solutions methodically and analytically until they find one that works. Far too many companies rely on their R&D departments as their primary source for innovation. But what this achieves, while it's not always a bad thing, is the development of a new process or technique for which designers have to find a use and marketers a user. For Avantis 2, we were able to articulate a concept at the very beginning of the process, initially as a series of sketches from the creative workshop and quickly thereafter as foam models, which the company's development team, as well as its senior management and marketing, could immediately get behind. Yes, many of us could see serious problems with it, but everyone could also see that it was compelling—that if we could get it to market at a competitive price, it would be a winner. After that, it became a question of how the hell to make it work and to manufacture it.

Following the Concept: The Aquaspeed

The concept we put in place was for what we called an "open-back" iron, in which the heel is completely open and separated from the body—a large loop on which the iron could sit for enhanced stability, but without adding bulk and weight (See Figure 4.21). Inside this loop, at the back of the iron, was a large trapdoor through which the reservoir might be filled more quickly and more conveniently. We suspected too that this new architecture might prove a useful solution to new and more rigorous EU-inspired "drop tests"—the loop potentially helping to absorb shock—and this proved to be true.

As the development team got to work in detail, we put aside some of the known problems and forged ahead as rapidly as possible to a finished model (See Figure 4.22—Figure 4.27). This was a change of process for us, as well as for the client. Previous to this project, we generally went through extensive development on each new product, absorbing considerable time and money establishing feasibility before that product could be researched with consumers, and then potentially be rejected. Better by far to gauge consumer interest in the broad concept as quickly as possible, which the model allowed us to do.

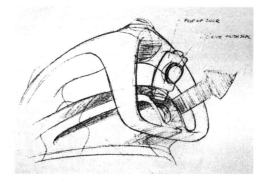

Figure 4.21 One of the earliest sketches for the "open-back" concept, which would make it possible to fill the iron with water via a trapdoor at the rear. Even at this very early stage, the whole idea of a light, open, airy structure of the heel (in order to reduce bulk) is considered in the context of allowing excess water to flow through and around the iron.

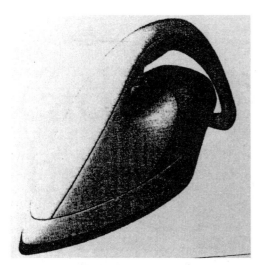

Figure 4.22 One of many early hand-made foam models that encapsulate the concept in three dimensions and allow more-detailed evaluation of the complex molding issues. The photo of the foam model has then been worked in Photoshop, coloring the tank blue and looking at alternative split-line positions.

Figure 4.23 The finished model, machined from Alias data. This model was crucial on a number of levels—it encapsulates the whole team's best thinking at a particular moment (even though much remains unresolved); it was used in consumer research; it became the focus for subsequent development; and it allowed pre-selling both within the business and outside.

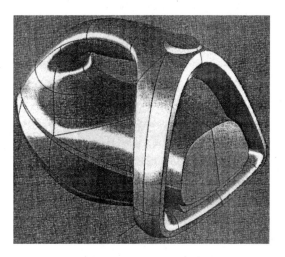

Figure 4.24 Defining the 3D form using Alias, the preferred software for resolving complex forms. These early surfaces were used to machine more accurate foam models to better reflect the technical probabilities and required volumes.

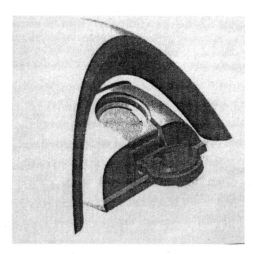

Figure 4.25 Detailed resolution of the trap-door-many different solutions had to be explored. Details like this ore often detached from the whole and studied piecemeal. Here, the designer is working in Pro-E looking at the implications of hinging low-down.

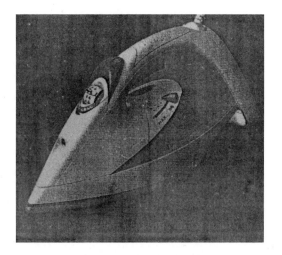

Figure 4.26 Finished photo-realistic rendering. Renderings and animations like this are a useful by-product of defining final surfaces in Alias. The primary purpose of finol surfaces is to allow the rapid production of a finished model.

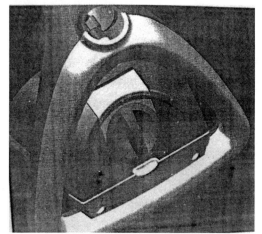

Figure 4.27 Getting closer to the final resolution-Calor's engineers are working on specific problems alongside Seymourpowell's designers. This is a screenshot from their Unigraphics CAD system.

  The Calor team's problems (only some of which had been anticipated) were just beginning—and this project would no doubt have failed in a company with a lesser culture of innovation than that of Calor. Crucial to its success was the vision, decisiveness, and guiding hand of CEO Jean-Pierre Lefevre, who has an instinctive feel for his products and their market. In my time, I've seen many great ideas at other companies become diluted, changed beyond recognition, or abandoned completely in the face of demanding technical challenges. But this CEO ensures that his team does not lose focus on the important issues.

Not surprisingly, most of the problems revolved around the innovation of filling the iron from the rear and had to do with sealing, air pockets, and venting, and with keeping water safely away from the electric components. Some of these problems required design changes from our side, particularly in optimizing the filling angle, but none required great compromise. It's a testament to the creativity of the development team that their work in finding solutions yielded new patents, which will make it difficult for Calor's competitors to catch up.

The new Avantis was eventually christened the Aquaspeed (See Figure 4.28). It was launched last January, and since then sales have more than met expectations—not just because this is a better innovative product, but because its design effectively communicates its benefits-speedy, fuss-free filling and improved heel stability without weight. In short, this iron sells itself. Aquaspeed is a credit to every member of the Calor and Seymourpowell team—and to the company itself. The Calor culture understands that innovation is the best, perhaps the only, way to maintain a strong brand in the face of low-cost "me-too" OEM products from the Far East. Of course, as a manufacturer of consumer products, Calor understands the fundamental role of design as a creator of attractive products. But much more important, it values design at a strategic level and as a creative catalyst for innovation... and that's why Calor is my favorite client!

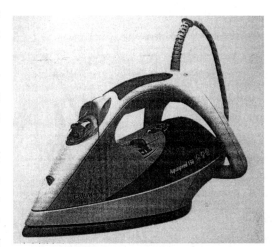

Figure 4.28 Aqunspeed—the finished product. Compare this with the finished model and (other than color) spot the differences: a "blink test" would say they are the same; a closer inspection would reveal some less obvious changes. Their similarity illustrates the close working relationship, respect, and understanding between Color and Seymourpowell—which, for anyone buying design services, is something to look for when assessing technical competence.

Free Reading 5

Formal Flexibility

BMW Gina Light

Over a year ago, in an interview with Auto & Design, BMW design chief Chris Bangle described his vision of the automotive philosophy of the future in decidedly abstract terms. "The important thing", he said, "is to start questioning dogmas: and sometimes, you need a big hammer to break them down. You just need to grasp the fact that everything is variable and that there are no fixed forms, so you can take a more flexible approach." It is only now, however, with the presentation of the Gina Light—a revolutionary experimental car skinned not with mental, but with a silver coloured elastic fabric—that this concept has taken tangible form. Even then, Bangle was already using the term Gina: an acronym standing for "geometry in an infinite number n of alternatives". In Bangle's mind, Gina was already a reality.

Saying that Gina Light is a "fabric car", however, would be too simplistic. While the use of fabric—a very special, high tech fabric, to be precise—is one of the possibilities opened up by this new philosophy, it is not the only one. The fabric skin of this model is only a means of visualising the variability of future car design and, therefore, of a new formal language. This iteration of Gina allows the creation of products with forms and functions that vary in relation to individual interpretation, to cater for the different demands of future users. As Bangle puts it, "the Gina philosophy, in its short form, is about being flexible—thinking flexible, acting flexible—context over dogma, that's it."

"First of all, we asked the question what do we need the skin of a car for anyway?", explains Bangle. "What's it there for, does it have to be made out of mental? In reality, the aspects of crash and stiffness and ride handling can be handled by a space frame type vehicle, entirely without the skin. And therefore to go away from the metal skin and adopt a skin that offers different capabilities, can be lighter and use materials that require less energy to make. Like the Gina concept, the possibilities are n-infinite." Within the company, Gina transformed form being a model to a shape and eventually become a philosophy and a thinking process. "Let's let materials talk in a different manner", says Bangle, "For instance, to get an idea of what the sculptural form is between the fender and, let's say, a wheel arch, you really only need a line around the wheel, and a line more or less where you want the fender to be, and you let the material do the taking in between." Seeing the door of the Gina Light bend outward, with its fabric skin wrinkling up, is actually quite alarming at first. But why shouldn't it? Just like a tailor made suit, you try it out a few times, than you make up the final one and put it on (and in the case of the Gina Light, the "suit" is donned in about two hours). In truth, this is actually a method already adopted by BMW for some others of its recent models. "Emotion is really the added value to this", continues Bangle. "And then there is the level of humanistic content that we can bring with Gina is about human way of designing cars."

Do we need to redefine the common concept of mobility? Do we need to rethink automotive construction? Is there a symbolic relationship between material, function and aesthetics? Which technologies will move us into the future? Can cars be like a tailor-made suit? The certainty, according to BMW, is that the relationship between humans and their cars will change, that functions will have an emotional impact. We need to question everything: for instance, does the roof always have to sit on pillars and be delimited by the windows? Are we certain that side by side is the ideal seat configuration? Do all functions really need to be visible all the time, even when not in use? A crucial postulation, however, is that the potential offered by new materials must find a place in the creative process of design.

According to Bangle, it is only by finding answers to these questions and abandoning dogma that a new form language will be defined enriches the aesthetics and expressiveness of a design process reflecting the individuality of the product, while taking production costs into account. The possibilities are endless, and BMW has already explored some of the avenues offered, such as the construct between convex and concave surfaces used in number of its production models or, in the case of the 2006 Mille Miglia concept, the application of solutions echoing the ancient Japanese art of Origami in the cockpit.

While we are still years behind NASA, which is developing new polymeric fibres that bend and change shape when crossed by electrical current and new memory alloys, the results could be very similar. "Demand for individualisation from customers", notes Bangle, "will introduce variety to the context of our products and change the whole image of the sector." The Gina philosophy will allow designers and research and production engineers to analyse the validity of current principles and processes, and examine possibilities open for cars of the future without having to comply with any conventional criteria, applying numerous different interpretations in the same concept. And this, if necessary, "may even mean questioning certain solutions considered to be beyond dispute tody". Some pictures of the Gina are as follows (See Figure 4.29—Figure 4.31):

Figure 4.29 Seem from the front and with no tactile interaction, the Gina Light looks just like a conventional car. One that offers infinite possibilities for modification, however

Figure 4.30 The sequence of pictures above illustrates the aperture of the non-existent bonnet. The Gina philosophy abandons numerous conventional automotive canons.

Figure 4.31 Left, a sequence of the doors opening

Core Text 12

Design Management

Definition of Design Management

◆ The Definition of Design Management varies with different people. But its essence is similar. Design management is the function of defining a design problem, finding the most suitable designer, and making it possible for him to solve it on time and within an agreed budget.

—By Michael Farr

◆ I define design as a plan to make something which in a management context is the cooperation achievement of product purpose, and also perhaps information about the purpose, design management, therefore, is an aspect of the planning process, through which organizations are run design management should be the central and vital aspect of the design process.

—By Peter Grob

◆ Design management is an interactive process involving the manager, those being managed, the activity being managed, the situation and the environment.

—By Oakly

Importance of Design Management

Design is now well recognized as a powerful tool in the hands of managers who need to make and sell products or develop an effective working environment. More and more business schools are adding design management to their curricula. The only remaining question is whether design as a key management preoccupation will continue to flourish. Its importance is more obviously in many aspects:

◆ It can help designers to develop their products with high qualities.
◆ It can help companies to use design as a resource in efficiently way.
◆ It can make design as a tool for reaching the goal of a company.
◆ It can help companies to set up the clean identity.
◆ It can help companies to communicate efficiently with consumers.
◆ It is possible for many senior executives now expecting to use design resources in the development and of business strategies.

Contents of Design Management

Managing general remains predominantly an art rather than a science, therefore, design management must always involve flexibility and judgment. It has a series of multiple meaning covering various levels of newly emerging design-related activities.

The work of these activities needs to describe:

◆ Design office management

This deals with the problems of managing a design practice, a consultancy or an in-house design department.

◆ Educating designer for management

This means equipping designers with the languages, and at least a nodding acquaintance with the norms and values, of the world of management in which they work.

◆ Educating managers to design

It is an important starting point for a satisfactory relationship between managers and designers. The task is to tear down those educational and cultural barriers.

◆ Design project management

This is about the place which design occupies in the project management process, when the process is the way in which action tasks are arranged in organizations. This is a central and vital activity, which design occupying the key role between the creative, innovation activities and the control of the preparatory stages of the operational tasks of the business.

◆ Design management organization

This is about the place that design occupies in the management structure of organization and about the variations and modulations that are needed to make that relationship effective.

Scope of Design Management

The management of design can be mapped out at two interrelated levels: the corporate design management level and the design project management level. At the lower project management level, the issues derive essentially from the shorter term, relatively confined problems encountered during the administration of design projects. In corporate design management, the issues center on the longer term implications of the relationship between an organization and its environment, and the contribution that design skills and activities make to this relationship.

The key issues encompassed by design management:

◆ At the corporate level
- The contribution of design skills to corporate profitability.
- Design policy and strategy formulation.
- Design responsibility and leadership.
- Devising and introducing corporate design management systems.
- Establishing and maintaining corporate design management systems.
- Positioning and intergrading the design resource within organizations.
- Corporate design and design management audits.
- Sources of new design investment opportunities.
- The legal dimension of design.
- Evaluating major design investment decisions.
- Design management development programs.
- Design and the manifestation of corporate identity.

◆ At the project level

- The nature of the design process and different types of design project.
- Design project proposals and the briefing process.
- Selection of designers.
- Bringing together and managing design project teams.
- Planning and administering design projects.
- Costing design work and drawing up budgets.
- Project documentation and control systems.
- Design research.
- Presentation of design proposals.
- Implementation of design solutions.
- Evaluation of design projects.

Essential Features of a Corporate Approach to Design

The essence of a corporate approach to design can be summarized as follows:

◆ Recognition that design represents an identifiable set of activities with an important contribution to long-term profitability, and thus needs to be managed rigorously.

◆ The professional management of design activities involves the integration of, and uplifting standard in all categories of design.

Key Words

[1] curricula [kəˈrɪkjʊlə] *n*. 课程

[2] in-house [ˈɪnhaʊs] *adj*. 内部的，机构内部的；在机构内部进行的 *adv*. 在机构内部

[3] modulation [ˌmɔdjʊˈleɪʃən] *n*. 调制

[4] manifestation [ˌmænɪfeˈsteɪʃən] *n*. 显示，表现，表明；表示

[5] preoccupation [prɪˌɔkjʊˈpeɪʃən] *n*. 1. 全神贯注，入神 2. 当务之急；使人全神贯注的事物

[6] design research 设计研究

[7] design management 设计管理

Key Sentences

1. Design project management

This is about the place which design occupies in the project management process, when the process is the way in which action tasks are arranged in organizations. This is a central and vital activity, which design occupying the key role between the creative, innovation activities and the control of the preparatory stages of the operational tasks of the business.

设计项目管理：

这是在项目管理流程中开展设计工作的一个阶段，这个阶段的主要方法是对行为任务

进行组织策划。这是一项核心和关键的活动，设计工作在创新性活动和商业操作任务准备阶段的控制中扮演关键的角色。

2. The management of design can be mapped out at two interrelated levels: the corporate design management level and the design project management level.

设计管理可被详细分为两个相互关联的层次：企业设计管理层和设计项目管理层。

课文翻译

设计管理

设计管理的定义

◆ 设计管理的定义对不同的人是有区别的，但是其本质是相似的。设计管理的功能是界定设计的问题，寻找最适合的设计师，让设计师在认可的预算内按时解决问题。（——迈克尔·法尔）

◆ 我把设计定义为一种计划，从管理的角度而言，它使得产品的目的以及包括这个目的的信息能以整体合作的方式完成。因此，我认为设计管理是计划过程的一个方面，通告计划过程组织采用设计管理使其成为设计过程中心且重要部分。（——彼得·格罗布）

◆ 设计管理是一个交互过程，这个过程包含了管理者、管理对象、管理行为、管理情形和所处的环境。（——奥克利）

设计管理的重要性

对于那些需要制作并且销售产品或者开发有效率的工作环境的管理者来说，设计目前被认为是他们掌握的有力工具，越来越多的商业学校将设计管理加入他们的课程中。唯一存在的问题是无论设计是否作为关键管理的当务之急，它都将继续蓬勃发展。它的重要性在许多方面都更明显：

◆ 帮助设计师开发高质量的产品。
◆ 帮助公司以有效的方式使用设计资源。
◆ 能够将设计作为一种实现公司目标的工具。
◆ 能够帮助公司建立清晰的识别。
◆ 能够帮助公司与客户进行有效的交流。
◆ 有可能许多高级管理者现在希望在开发与商业战略中使用设计。

管理总的来说，首先是一门艺术，其次才是一门科学，因此，设计管理必须经常融入灵活性和判别力。它有一系列多样的意义，这些意义覆盖了新出现的与设计相关活动的各个层面。

这些活动的工作描述如下：

◆ 设计办公管理

处理管理实践，咨询或内部设计部门的问题。

◆ 培训设计师管理知识

就是说设计师在其行业内学习管理领域的一些知识，至少学习规范和价值观等方面的一些浅显的知识。

◆ 培训管理者设计知识

这是管理者和设计师良好关系的重要起点。其任务是打破那些教育和文化障碍。

◆ 设计项目管理

有关设计在项目管理过程中所占据的地位，这个过程是行为任务参与组织的方式，这是中心并且重要的活动，设计在创造性的，革新性的活动之间占据关键地位，并且控制着商业操作任务的准备阶段。

◆ 设计组织管理

这是有关设计在组织结构管理中所占据的位置和有关需要使这种关系更有效的变量和模块。

设计管理的范围

设计管理可被详细分为两个相互关联的层次：企业设计管理层和设计项目管理层。在较初级的项目管理层，问题来自于在设计项目管理部门中遇到的较短的，相对有限的问题，更短的术语。在企业设计管理中，问题集中于长期的组织和环境之间关系的暗示和设计技能与活动对这种关系的贡献。

设计管理包括的关键问题如下：

◆ 企业层面
- 设计技巧对企业利益的贡献
- 设计策略和战略构想
- 设计责任和领导
- 设计和介绍企业管理系统
- 建立和保持企业管理系统
- 在组织内部定位和集合设计资源
- 企业设计和设计管理审计
- 新设计投资机会来源
- 设计的法律层面
- 评估主要的设计投资决定
- 设计管理开发程序
- 设计和企业识别表现

◆ 项目层面
- 设计过程的本质和设计项目的不同类型
- 设计项目计划和简要过程
- 设计师的选择
- 带领和管理设计项目团队
- 计划和实施设计项目
- 核算设计工作和起草预算
- 项目记录和控制系统
- 项目研究
- 设计计划的表现
- 设计解决方案的执行

- 设计项目的评估

企业设计的基本特征

企业设计的本质可总结如下：

◆ 认识到设计代表了可识别的系列活动并对长期收益有着重要的贡献，因此，设计需要进行严格管理。

◆ 设计工作的专业化管理包含了设计一体化管理和设计所有范畴中令人振奋的标准。

Free Reading 1

Integrated Branding

Like it or not, your organization and the products or services it sells, have a brand. It is the sum of all the impressions your prospects and customers collect from the first time they hear your voice, see your brochure, or link to your web site. And if you don't take branding seriously, you're leaving a critical piece of the marketing puzzle to little more than chance.

Establish the Idea behind the Brand

Advertising pioneer David Ogilvy referred to a brand as a "product's personality its name, its packaging, its price, the style of its advertising, and above all, the nature of the product itself." How important is your personality to your everyday life? That's how important your brand is to your business.

The brand is less about your organization than it is about the product or service it offers. Customers buy a product or service because it offers a benefit—it solves a problem, it saves money or time, it supports their attitudes or beliefs, it is pleasing to their sense, and so on. They favor a particular company because it offers the best price on a widely available product, it provides better service, has a superior reputation, and so on.

The first step in creating a new brand, or fleshing out an existing one, is to define those benefits. They should be the very essence of your organization—the foundational elements of every marketing effort and advertising campaign. Defining those benefits is the conceptual side of branding, but I want to focus on the other side-the visual side.

Develop a Visual Palette

I call the visual elements we use to present those brand ideas, a visual palette. It includes all of the basic components you use to design most—a logo, typefaces, artwork, photographs, and color. Combined, they equal an image that distinguishes your organization from all others. Once it's established, everyone involved can use it to build a brand that is both unique and consistent.

Should you create your own visual palette? If you're not a designer, the question is an important one. I'm a big believer in doing only as much as you are comfortable doing. If, for example, you feel the type of talent you can afford to hire couldn't possibly do a better job than you could do yourself, by all means, develop your palette yourself. If, on the other hand, you aren't comfortable designing your own logo, choosing typefaces that work well together, or picking a palette of colors, and don't want to learn, pass the pieces you are not comfortable with to a pro.

Remember this: developing a compelling message and a strong visual palette is not the place to skimp on time or money. I've seen countless cases of companies willing to invest tens of thousands of dollars for printing, ad space, and the sales staff to publicize a brand they spend next to nothing to create.

Start with the Result in Mind

Start by deciding how you want people to see your product, service, or cause. An outdoor outfitter, for example, wants an entirely different image than a bookkeeping firm-a natural, relaxed attitude versus a buttoned up, highly organized one. Study the brands being developed by your competitors. Read their advertising and marketing materials, visit their web sites and those of similar businesses in other parts of the country to see how they distinguish themselves.

Remember, focus on branding your product or service, not your company. By that, I mean a company selling turn-of-the-century furniture reproductions may have a technologically advanced manufacturing facility and a progressive management structure, but its message and its image should focus on that turn-of-the-century style.

If you have drastically different types of products and services, do what the big guys do—develop a different brand for each. I'd venture to say we all know more about the individual brands of Doritos and Tropicana Orange Juice than we do about their parent PepsiCo.

Assemble the Pieces

Start with a logo and a display typeface. Typically, your logo is the foundational element on which you build your visual palette. If it is included on your signs, product packaging, brochures, stationery, and such, it stands to reason that it should be the visual center of gravity.

Once I had the logo in the pace I matched a typeface to it. Combine it with an illustration style. The best artwork and photographs express something words alone cannot—they establish a mood, explain your idea, demonstration a benefit, or show people, places, and products.

Choose Colors

Last you should select a combination of colors to use throughout the clients materials. The same selection process applies to any palette—design the logo first, choose the primary typeface second, select a collection of photographs, an assortment of clip art images, and last, choose two or three basic colors.

Create a Palette and Stick with It

If your message and visual style are working, stick with it. Too often clients get bored with a long-standing brand or new players make change for the sake of change. Though you may see your brand every day, remember that your prospects and customers do not. They need to hear, read, and see a consistent message over a long period of time for your brand to have maximum effect.

Free Reading 2

Forgotten Bond: Brand Identity and Product Design

The question is: How can the design of products communicate brand values? Guido Stompff distills answers from a case study about Océ Technologies, a Netherlands-based document reproduction and management company. His analysis reveals how designers, as members of the product development team, exploit the profile, proportions, interface, and even the color of Océ machines to convey and maintain a reputation for reliable professional equipment.

What do we talk about when we talk about branding? Logos, usually—and brand names, packaging, interactive branding, and advertising. There are frequent references to products, but products are always the subject of a brand, never its object! Still, products, the most basic marketing tools, do communicate—and not just through their function or through the way they are used. They arouse emotions, create experiences, age, and break down. Identity and brand are strongly influenced by the emotional response of people who use the products. This article discusses the way a product elicits emotions and thus affects brand image. It will then show—through a case study—how brand values can be translated into product design.

A Theoretical Framework: Products and Emotions

We all have products we like to use every day—even though they may be worn out—and occasionally we find ourselves buying things for which we have no actual need. Why do we believe that a certain product is better than all the others, even though we have never tested the alternatives?

It is evident that products (and brands) have an irrational, emotional appeal, but is it possible to manipulate this appeal, to design products that will contribute toward a desired brand image? Let's get to the basics: the relationships among people, their emotions, and their products, expressed as concerns, standards, attitudes, objects, and agents (See Figure 4.32).

Concerns are the more-or-less-stable personal preferences of an individual. These can be goals one wants to accomplish, either basic ("I want to eat") or more elaborate ("I want to make the school debating team"). Concerns can also be standards, in the sense that they are beliefs about the way things should be, such as "Products should be simple to use." Concerns can also be attitudes-for instance, "I'm crazy about B movies." Concerns are personal, but they can be shared among many of us. Consider such subjects as the benefits of democracy, the problem of environmental degradation, and so on.

And here's where product design comes in. Consider a product as a stimulus. It can be viewed as simply an object ("What a beautiful chair!"), but just as easily as an agent that represents something else: a company, a social group, a beloved one. People tend to use these products because of the sense of belonging they give (let's face it, no one buys a

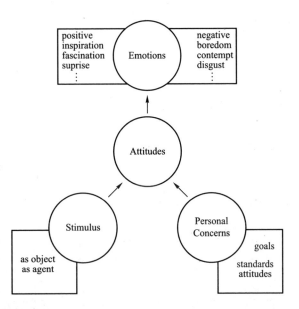

Figure 4.32　This model, taken from paul Desmet's Thesis (see Suggested Readings), shows the relationships among people, their emotions, and the products they purchase

Rolex just because it is reliable).

Emotions can be considered as mechanisms that signal when events, things, or persons are favorable or harmful to our concerns. Emotions related to products can be either pleasant (for instance, desire or amusement) or unpleasant (disgust or boredom). All emotions are preceded and elicited by appraisal—the unintellectual and direct judgment of a stimulus for its effect on one's well-being.

Because they stem from the concerns of an individual, emotions can differ quite a bit from person to person. Indeed, occasionally they are mixed, since it's quite possible to have concerns that contradict each other.

Brand, Products, and Emotions

What is the relevance of this framework in relation to branding? Emotions signal when events, things, or persons are favorable or harmful to our concerns. These emotions surely contribute to a brand's identity. However, it is hazardous to claim that a design will generate the emotions a designer or marketer wants, and thus create a certain brand identity. Concerns are individual and constantly changing, and consequently so are emotions. Emotions generated by a product can be numerous, not only between individuals but also within a single person. End of the story, then? No. This framework does afford some interesting starting points for designers and brand managers in understanding how products can contribute toward a desired brand image. For example:

Which client concerns are universal?

Is the product appraised positively as an object, or does it actually represent something else?

Which design features contribute toward this favorable emotion?

Case Study: Océ

Océ, the 125-year-old international company for which I am a product designer, has its headquarters in The Netherlands. It changed its core business several times before limiting itself to the development, production, and supply of services, software, and equipment for the reproduction and management of documents (See Figure 4.33). It is a business-to-business company, entirely oriented toward large clients—companies, government authorities, universities, and so on. The company develops professional products for the office market and is the leader of the world market for engineering/CAD environments and for the high-speed printing market. The company has a strong reputation due to its brand values—reliability, productivity, and user-friendliness.

Figure 4.33 Océ product

It is always hard to pinpoint a corporate culture, but Océ's could be described as nonhierarchical, responsible, truly customer-oriented, and a bit nonconformist. These values can be seen in, for instance, the color engine at the core of its imaging technology products. Instead of superimposing the classic four colors to produce all the others, the engine uses seven colors arranged in adjoining dot patterns. (I can hear readers thinking, "So what?" but these differences were not solely invented to keep R&D busy. They confer several advantages, such as the ability to use heavier papers or specially coated papers, or the ability to fold prints without cracking or flaking thick layers of toner.) In short, Océ lacks the size of a Xerox or a Canon, but it still chooses its own direction, because it feels this serves its clients better. This essentially sums up our corporate culture.

Concern: Are You Professional?

Understanding the concerns of clients can provide insight into their preferences for specific brand values, resulting in a product design that communicates these values. Companies and organizations do not normally consider their document management as a core business activity; consequently, they look for partners who can provide adequate solutions. The main users of Océ products are equipment operators, whose greatest concern is that their equipment be reliable. International test reports frequently refer to an Océ product with remarks such as "a sturdy workhorse", "looks like a tank, built like a tank", or "very plain-looking; simple interface" —not comments designers normally enjoy hearing! However, Océ has won more than 40 design awards since we started counting, about 10 years ago. Apparently, the company's designers are able to create products that communicate an important brand value—reliability—without relinquishing quality in product design.

But a reliable appearance is not enough. Océ clients also expect a professional attitude from their business partner. And that implies that the company presentation needs to be professional, as well-suffusing not just the brochures, showrooms, and Web site, but the products too, of course. There are a few rules of thumb here. First of all, "family resemblance" is preferred above the distinctive appearance of individual products. That is, you buy "an Océ", not "a 3165", just as you might buy "a BMW", "a Sony", or "an Apple". Second, the best way to look professional is to choose an independent design policy instead of following trends. These explicit design choices even gained importance during the last decade. Océ products were originally stand-alone products—that is, they were copiers, and only copiers. Nowadays, solutions for Océ's clients include several printers, scanners, copiers, and applications. Combined, they create an image of Océ, rather than of each product separately. Although the individual products were developed by several R&D establishments at Océ, together they have to present a well-considered total solution.

Innate Attitudes: The Things We All Like

Attitudes are a species of "concern" that involves personal preferences. Because attitudes are so individual, it is hard to use them as starting points for design. However, some attitudes seem to be innate and common to most human beings—for example, the affinity for symmetry. For instance, people whose appearance is very symmetrical are generally regarded to be more attractive than others.

Designers are well aware of these innate attitudes, and product design reflects this. Small cars tend to have relatively high bodies, large headlights, and small wheels in order to elicit nurturing, affectionate feelings. Babies, after all, have similar proportions. At Océ, the designers work with a carefully developed system of dimensions that are multiples or divisions of 12. In addition, the Océ designers use repetitive proportions in our products, from the smallest part to the entire product-symmetry in the classical Greek sense. (Although it is a great story with which to impress other designers, I have not yet spoken to anyone at Océ who has actually noticed this independently!) But Océ products are often described as "calm, restrained, tranquil, and balanced" —exactly what the Océ designers think is appropriate for professional products, equipment that is used all day long. People are well able to sense implicit order and to value it. Above all, this symmetrical inclination has proved to be a very strong instrument for producing an instantly recognizable family relation among different products—even when the development of those products happened 10 years apart! In a way, it is almost like Océ's DNA—an integral part of all the members of this company's product family (See Figure 4.34—Figure 4.36).

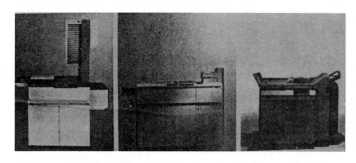
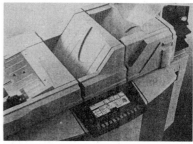

Figure 4.34 These successful Océ copiers were designed in different decades, but they were All Developed for the same market offices. They Show the company's very characteristic "brand DNA."
Top, from left to right: the Océ1900 (1980), the Océ 2465 (1988), and the Océ 3045 (1994). Below: the Océ varioprint 2070 (2002).

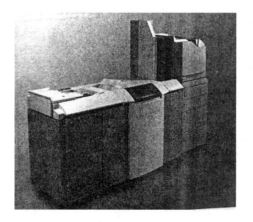

Figure 4.35 Integrating modules and frames into a strong contour

Figure 4.36 Océ's full-color graphical user Interface is designed for professional operators in central reprographic environments. It is used in several products targeted for this market with a focus upon workflow and productivity. The design, including typography, colors, and icons, reflects the colors, icons, and typography used in the physical product design, creating a truly professional feeling.

Learned Attitudes That Are Shared

Attitudes can be learned, in the same way that one acquires a taste for beer or red wine. Such attitudes are personal to individuals, but cultural influences can also lead to similarities in attitudes. Research we carried out concerning consumer perception of our products and competing products showed that noticeable "modularity" in products reduces their confidence significantly. People are well able to recognize products that are composed of different modules, and to mistrust them. The most likely reason is that we have all learned that products that are an assembly of modules simply are not as good as dedicated products.

Océ products do not look like an assembly of modules. Our designers favor a basic, strong contour with clear lines, into which all modules "snap" (See Figure 4.35). Indeed, in the case of CAD printers and scanners, where competing products are supported upon a tubular frame that has nothing to do with the design of the product itself, Océ products have a robust support that blends harmoniously with the actual printer or scanner.

Appraising Products as Such

Again, returning to the discussion of concerns and their different manifestations, consider that concerns can be experienced as goals or objectives, and that a product can be evaluated in terms of its use in accomplishing those goals. Thus, at Océ, functionality is an ever-present issue, as we determine which tasks our products should be able to perform. The digitalization of our products and the introduction of new software applications present a host of new possibilities, scaling in different directions, mirroring images, optimizing for photographs, and so on.

Interestingly, many—even most—of the new functions possible are never incorporated into our products, simply because our clients don't really need them. When a user copies a document carrying photographs, he or she prefers a product that recognizes the inclusion of photographs and automatically provides that optimization. Restraint requires endless discipline, but without it, gimmicks would otherwise clutter our interfaces and our image.

A product can be appraised for its usability. Does it intuit the promised functionality? How many actions or clicks are required? Is the result a logical one? People can be truly disappointed if they fail to access a product's functionality because the interaction is too complicated. Just consider the introduction of the handheld assistant for the Palm Pilot. It immediately made all its predecessors obsolete. Here is a prime example of excellent usability being recognized instantly.

Products can also be appraised for their aesthetics, for the way their appearance makes you feel. This is personal, but sometimes there are design features that arouse similar reactions in a variety of people. One Océ example is the use of color. Five years ago, we decided we needed to develop a new range of colors. At that time (to paraphrase Henry Ford), you could have office equipment in any color, as long as it was beige. The Océ designers decided to use a color range based on a fresh green and gray; it was tranquil, fresh, and sufficiently different from the light brown everyone else was using. Moreover, we designed the colors of our Graphical User Interface (See Figure 4.36) to fall in line with those product colors, thereby creating a coherent product feeling. It worked out better than anticipated; not only do the adjacent PCs look old-fashioned, but so do competitors' products that are, quite frequently, situated close to the Océ products. The Apple color revolution of the past few years has also been an unexpected benefit to us!

Appraising Products as Representative Agents

Some products just naturally lend themselves as agents that represent something—a designer, a company, or a social group. If your peer group is wearing jeans two sizes too large, you end up not only buying, but also really liking, those shapeless jeans! Copiers, printers, and scanners would not seem to lend themselves to this kind of thinking. They are professional products; you would expect that choosing them over another product would be mostly a matter of functionality. At least, that is what I thought until two years ago. At that time, Océ introduced a high-end product with a separate control station, designed to be used by someone who is either standing or sitting (See Figure 4.37). After the initial introduction phase, a visit to one of our clients showed us an unexpected side effect. They had placed the product near the entrance to their repro room, to show off their skills in using this high-tech device. The product served as an agent for the "IT worker" as opposed to the "engine operator." It is something to keep in mind—professionals enjoy being able to demonstrate their skills.

A product can also be an agent for a story. That is, when they purchase a particular product, people may really be buying the stories, legends, and emotions surrounding that product. If you want to feel free, you buy the symbol of freedom—a Harley-Davidson. Océ discovered that a story should be based on your true nature, your identity.

Océ products also signal a particular identity. For instance, while our print technology guarantees excellent print quality, it also means that our products are bulky—noticeably larger than competing products. Our designers dealt with this issue by using height. Océ modules are sometimes positioned on top of each other. This creates immediately recognizable product contours, as well as improved ergonomics. Of course, some stories work better with our products than others. Just as being "high-tech" is often associated with smaller products, being "independent" fits well with the larger, and recognizably contoured Océ products.

Designing Identity to Fit Culture

This case study may give the impression that brand values were the starting points for Océ design projects—that our designers "just knew", but the opposite is nearer the truth. The Océ designers are part of Océ R&D, where all our products are developed. They developed the company's design language through practical experience, experience that is constantly evolving and gradually adapting the brand's DNA. The design of both products and interfaces proved to be capable of communicating Océ brand values (See Figure 4.38).

This demonstrates one of the main conditions for design for brand identity—a long-term relation between the designers and the company. The designers need to understand the position of the company or brand. They need to breathe and absorb the culture, to feel "how we do things around here" —they need to become the brand. This can be achieved in two ways: through having an in-house design studio, or through a long-term relation with a designer or a design office. Using new designers for every new product can result in the creation of an array of beautifully designed, award-winning products. However, eventually, you risk losing brand identity—unless, of course, being trendy is one of your values.

Figure 4.37 The Océ DPS 400 offers a separate control station that can be used by a person in either a standing or a sitting position.

Figure 4.38 The Océ TCS 400 CAD color printer demonstrates that communicating great design and usability (embodied by a tiltable and rotatable interface) helps to create a favorable Brand image

Corporate design for products should be a direct translation of the culture for which a company or brand stands. Océ products developed in the early 1980s are still in use. Trying to create a brand image that is different from a product that is still in use will very probably be perceived as a superficial effort. It is better to "tell stories" that are in line with what you are, and strengthen these through designs. Consider Harley-Davidson again. Most likely, Harley's idea of what a motorcycle should be (and how it should sound) is what attracts Harley's fans, and that idea says: Don't be like the others. The story came later. When you choose to embed the design of products in the culture of a company, you never send the wrong message.

Summary

Products arouse emotions—occasionally even mixed emotions—that contribute strongly toward people's feelings about a specific brand. Emotions are individual, as are concerns, and they are always changing and developing. Therefore, it is hazardous to claim that a design will generate specific emotions and therefore create a certain brand identity. Nevertheless, there are interesting starting points. Find the common concerns of your customers. Understand why a product is appealing. Is it because of the "object" itself or is it an "agent" representing something else? Find out which design features evoke favorable emotions and use them to develop your own "brand DNA". Last but not least, product design should be rooted in the culture of a company to ensure a consequent message—because, if they're any good, products will outlast any brand identity campaign.

Perspectives on Designing Design Managers

Design management integrates two very different realities. To probe the background needed to embrace this duality, we invited six experts from the corporate, consultant, and academic worlds—Lee Green, Jeff Smith, Gary Bryant, Rachel Cooper, Kyung-won Chung, and Maryann Finiw—to comment on the business knowledge and attitudes that distinguish designers from design managers.

A Business Profile for Design Managers (Lee D. Green, Director of Corporate Identity and Design, IBM Corporation)

Today designers, and particularly design managers, are coming from increasingly diverse professional and educational backgrounds. This is in part due to the emphasis universities are placing on multidisciplinary team program structures. It also reflects many students' desires to expose themselves to a diverse field of studies prior to focusing on a specific career discipline. As a result, the field of design is increasingly chosen after study in communications, philosophy, anthropology, business, information technology, and other concentrations. This diversity, when combined with a high-quality formal design curriculum, is a great background for today's designers and design managers.

However, there are very few formal design management programs (undergraduate or graduate) offered by universities or design schools. This is unfortunate, because design management is a unique discipline and requires a unique set of learned skills and methodologies. In most companies, design managers evolve from the ranks of design practitioners. This is a logical career path. However, many practitioners struggle with the adjustment required to shift from the familiar "qualitative" approach, adopted to solve design problems, to the more "quantitative" approach required to influence business decisions. These new managers often passionately advocate the value of design, but too frequently "speak" in the language of designers... not marketing or business managers. To be most effective in leveraging what they instinctively know about design, they need to make an important transition. If they do, they can effect change. If not, they are viewed as implementers of tactics, not as strategic contributors, and this limits their impact and growth.

One way to address this challenge is for more university and design school programs to develop a curriculum that includes design management. I believe curriculum should focus on three components of a repeatable process.

The first is "problem analysis and engagement". This involves an understanding of market conditions and the competitive environment, as well as the ability to select and model related research initiatives. It also requires an understanding of how to engage the right "partners" in the organization and how to create buy-in and internal champions.

The second is "packaging the analysis and the proposal". The focus here is on techniques for building and articulating an influential story in the language of business. It also requires that the story be clearly linked to business strategy and goals. Most would agree this is a logical process, but it is not well understood by today's designers, and it is rarely taught. Nevertheless, these are the skills companies look for when hiring marketing professionals.

The third is "piloting and operationalizing". This involves demonstration of success and results through creation of pilots that can be tested and the related measurement to validate their effectiveness. Without this step it is very difficult to gain the internal commitment for funding and resources required to support broad adoption.

In complex matrix organizations, these are all steps that require management and negotiation skills, as well as business savvy, and of course, expert knowledge of design. But expert knowledge of design alone is rarely sufficient to allow for successful navigation through the complexities required to influence change in an organization.

The burden to teach these competencies and practices does not fall entirely on the universities or schools. Companies need to cultivate these skills within their organizations, and professional organizations, like DMI, need to continue to provide advanced education and related forums.

The next generation of design managers is up to the challenge. They have the desire, energy, intellect, and inspiration. They are ready to utilize their diverse capabilities and talents to become agents of change. Let's help them.

Value for Design and Design for Value (Jeff Smith, CEO, LUNAR Design)

In the interest of creating shareholder value, business leaders and managers are generally focused on two primary and very pragmatic goals: grow revenues and increase profitability. Business leaders and managers are most interested in design when it helps create financial value. Their orientation is "design for value".

Designers' goals are typically individualistic or idealistic: being original, for instance, or standing out in style and fashion, helping others live in a better way, or doing their part to save the world. Designers often value design for its own sake. Their orientation is "value for design".

Bringing these two orientations together is a challenge that is becoming more valuable as more businesses experience higher levels of consumer-driven competition. Chris Zook, a director at Bain & Co. and the leader of its global strategy practice, conducted research for 10 years, looking at 100 companies worldwide to reveal the secrets of successful and consistent growth in financial value.

He discovered that the top 25 performers—companies such as Nike, P&G, Hilti, and Legend—had two things in common: They built their growth around and out of their core markets and they gained insights into how to grow by getting very close to their customers. A good example of this can be found in the teeth-whitening strips and low-cost motorized toothbrushes that P&G markets to broaden its business around conventional toothbrushes. The motorized toothbrush was invented by industrial designers who transferred to the toothbrush market the know-how they had accumulated through designing spinning lollipops.

Designers love to run from the status quo to the new and original. They are also very good at inventing and creating solutions for people. To bring more impact to our client's businesses, we look for three types of designers: 1) inventive and creative designers who understand how to use their originality to help our clients grow from their core business markets; 2) designers who excel at understanding people and who can invent and create new ways to improve their lives; and 3) designers who can leverage design to increase the quality of products while making them less expensive to manufacture.

In short, our goal is to find designers who value design but know how to design for value.

Training Tomorrow's Design Managers(Gary Bryant, Manager of Industrial Design, Caterpillar Inc)

Looking back over our respective careers, most of us would likely say, "I wish I had known then what I know now." It must be a universal sentiment in most fields of endeavor, but I often consider how different a designer I would be today if I could only have reached into my pocket and read through a few notes from the future me. Oh, well—so it goes in time travel and daydreams, but I would submit that each of us could share "lessons learned" with younger designers or new members of our team in order to speed understanding or avoid some of the bumps in the road to experience.

This would serve our customers in many ways, one of which would be the potential for increased customer reliance on design as a business resource. Increasingly, designers must address or learn aspects of total project management, sourcing, purchasing, quality assurance, accounting, and return on investment, among others. As customers look for more "turnkey" or "total" solutions from design firms, some business functions not covered in core design curricula must be learned somehow. Many design schools and design firms are addressing these added functions, but the pace must be quickened.

At Caterpillar, the enterprise-wide infusion of Six Sigma methodologies has substantially changed the expectations put on each employee, including, of course, designers. A deeper understanding of economics, marketing, finance, organizational management, and strategy is needed at various levels of our projects. Ongoing training within the company is critical, but new employees, as well as current students, need more insight into these areas. Local colleges and universities would do well to consider their corporate neighbors when crafting supplemental training programs that focus on business functions and are targeted toward designers and technical types. A local university near Caterpillar's world headquarters has in recent years done just that and put together a comprehensive 16-week business course for technical professionals. The results are excellent, and such a program shows our customers that all aspects of their business are important to the design team.

When considering design talent, we look to add those individuals who are great designers, but who also understand customer business models. We share with them as quickly as possible what we feel they need to know, but we are also willing to get them training in areas that cannot be addressed internally. The learning curve is steeper and more costly when done internally, but it can be optimized to fit the needs of the enterprise and the individual designer.

In light of the increased expectations put on design teams, the business understanding thus acquired will build a trust in and a reliance on designers that transcends traditional capabilities. It will also bring design into the customer's thought processes—which is, after all, key to any successful product design solution. In the long run, maybe it will also make that "lessons learned" note we leave for ourselves from some future date... just a little shorter.

The Design Leader and the New Designer(Rachel Cooper, Professor of Design Management, University of Salford)

After 20 years of researching the relationship between the designer and the business, the one thing I am sure of is that there are no simple solutions to making design an effective organizational resource. Every time we study a designer or design manager and his or her corporate role, we find new insights. Take, for example, two companies—one a manufacturer of sportswear, the other a financial services firm—in which we investigated the role of the design manager and you will find two design managers with completely different roles.

For the sportswear manufacturer, the design manager's role is to provide design direction and focus, manage a team of designers, and provide an interface with other corporate activities. The key skills for these tasks are design understanding, design leadership, and personnel management; hence this design manager could be said to be a creative team manager.

The design manager at the financial services firm spends most of her time managing a relationship with external design consultants, defining corporate design policy, and negotiating solutions. Her key skills include design awareness, process management, and a general knowledge of the financial services business. This person could be said to be primarily a design procurement manager.

Looking at these and other situations in which design clearly contributes to company success, we frequently find that the designer (either internal designer or consultant designer) has worked by stealth. He or she has used personality, as well as communication skills both visual and verbal, to learn about the organization and to access crucial knowledge and understanding—not only within the company but also from its supply chain. Thus, the design manager needs to have a significant amount of authority and access to people within the organization, as well as an openness to change and new ideas. A good understanding of business issues is also crucial. And there is one other necessary ingredient: an intuitive or empathetic feeling for the essence of design.

Successful design managers often become design champions within their organizations. They tend to be empathetic; they are themselves designers or have some characteristics of the "design mind"; they are good networkers and communicators—particularly of the organization's brand values—and they must have the ability to facilitate and manage relationships with users and suppliers. For the organization's part, it must define the roles of design champion, design manager, and designer, yet be flexible and dynamic enough to enable them to drive innovation. It is clear, then, that organizations must understand how design contributes to its overall strategy and how it adds value to its businesses.

The design champion can form his or her own design team or be supported by an existing team. This is a critical task, and the components of a team will be driven by the needs of the business. For instance, the company may decide to recruit an in-house designer or retain a design consultancy with an eye to establishing a good rapport and understanding of the business and its design needs over the long term. If the company determines a need for radical change or innovation, it may select a designer who has no previous experience with the business or indeed the sector—someone who can act as a "disruptive influence" and bring in new ideas. British Airways took this approach when they needed to re-brand their club-class offering; they asked their internal design managers to select an external design consultant who had never previously designed for aircraft. This design group, Tangerine, challenged conventional design and regulations to develop the first seat bed, and enabled British Airways to gain market share back for club-class seats.

Research suggests that the role of the design manager is in constant flux. In our latest book, *The Design Experience* (Ashgate Publishers, 2003), Mike Press and I describe the future design professional working in sustainable economies, by which we mean international, national, and corporate environments that are concerned with economic, social, and environmental issues. These design professionals will be intelligent "makers" (people with craft skills who are also creative problem solvers and reflective thinkers); active learners and communicators (skilled users of networks); active citizens (internationalist, empathetic social initiators); and sustainable entrepreneurs (ecologists, business strategists, managers, and marketers). The new design champions will need to be comfortable in all these dimensions, and design education will also need to change if they are to learn the language of design with an enhanced knowledge of the social, political, economic, environmental, and technological context.

Searching for a Creative Generalist(Kyung-won Chung, Professor of Industrial Design, KAIST)

As companies increasingly recognize design's contributions to business success, the position of design manager has become more and more important. Although design managers may be given a range of titles—chief design officer, design director, and so on—depending on the nature of their tasks and the corporate environment itself, the major responsibility for all of them is to solve whatever design problems and difficulties arise. Beyond that, however, there are several more tasks and qualifications that are specific to the design manager.

The first of these tasks is that of enhancing corporate value through the strategic use of design. The development of a corporate design program for establishing high standards of corporate image and brand identity is one aspect of this function, as is the creation of a corporate design agenda.

Another task is the proper management of the design organization. This includes establishing a flexible organizational structure that can reconfigure the resources necessary to cope with emerging opportunities. It also involves recruiting and developing designers to play specific and significant roles. Providing comprehensive programs for the renewal of design skills and enlarging the information and knowledge base of designers is also important, as is the care and maintenance of sophisticated design equipment.

The third task in the effective management of design projects is to integrate design into the entire product innovation process. To do this, the design manager must maintain a highly creative environment—one that allows designers to be more sensitive, imaginative, and intuitive. Since creativity cannot be encouraged under a pattern of strict routine or exhaustive regimentation, the design manager should try to eliminate the kinds of obstacles that hinder creative thinking. Establishing a highly effective collaboration network between the design project team members and others is also very important.

Design managers must have some expertise in manufacturing, as well as a comprehensive understanding of people and their motivations, design processes and methods, potential markets, available technologies, and so forth. However, it is very important to have the capability to solve problems creatively and to be able to call on excellent aesthetic judgment.

We can quibble about the various qualifications of the ideal design manager, but one thing is certain: he or she must be a generalist who can integrate a variety of design issues, not a specialist with somewhat specific skills. A design manager should know everything about design and something about all the related disciplines-especially management. In other words, he or she should have a broad range of knowledge and the capability for effective communication with others. If a designer limits himself or herself to a narrow technical corner, he or she will remain a specialist.

And not every senior designer can become a good design manager. The skills required include the ability to see the "big picture" and a grounding in the knowledge and skills of management, as well as design. A young designer who aspires to the position of design manager should try to think as a manager in a leadership position would; this is a good way in which to accumulate experience, however indirect.

Training and Experience in Business (Maryann Finiw, Principal, Strategy Practice, Design Continuum)

The most attractive job candidates are the ones who are motivated to apply their design talents and skills to meeting business objectives. Ideally, this means that they describe a project as successful not just because it won a design award, but also because it met a business objective. I like to hear candidates talk about increasing market share, improving margins, and reaching new market segments as their key metrics for project success. This demonstrates that they are ready to solve business problems through design.

Some of the most successful candidates are those with varied backgrounds that allow them to contribute multiple perspectives on design and business issues.

Designers need to understand the business perspectives on design issues. Not all clients are product development managers, which mean that many design decisions are influenced by people from marketing, market research, brand management, engineering, or other disciplines. In order to succeed, designers need to learn to speak their languages and to gain insight into their decision-making processes.

Design Continuum realized this early on and developed training programs in Marketing 101, Market Research 101, and Clientopia. These training sessions help our designers place their projects in a business context and create a common language for communicating about them with clients. Marketing 101 covers the basic marketing P's: people, product, price, place, promotion, and positioning. The goal is to recognize that the center of our universe is not necessarily the center of the client's universe, and to understand what pressures and priorities the client is facing.

Market Research 101 offers common tools for making business decisions. For instance, it teaches designers how to intelligently analyze research provided by clients and builds their skills in advising clients on what types of research might be most appropriate at different stages of the design development process.

For Clientopia training, we invited a long-term client to discuss our projects from her perspective. She described how our projects fit into her company's other initiatives, compared with her other job responsibilities, and what aspects of our project process and deliverables were valuable to her and why.

This training has resulted in behavioral changes. Our designers and engineers now request to review their clients' competitive positioning and target market research. They have learned to ask the client questions that get at all the marketing P's. This makes us more valuable to our clients in solving their business problems through design.

Core Text 13

IDEO

IDEO is a design and innovation consultancy based in Palo Alto, California, United States with other offices in San Francisco, Chicago, New York, Boston, London, Munich and Shanghai. The company helps design products, services, environments, and digital experiences. Additionally, the company has become increasingly involved in management consulting.

IDEO was formed in 1991 by a merger of three established design firms: David Kelley Design (founded by David Kelley, who is also a professor at Stanford University), ID Two (founded by Britain's Bill Moggridge), and Matrix Product Design (founded by Mike Nuttall). Office-furniture maker Steelcase used to own a majority stake in the firm, but is spinning out the subsidiary through a five-year management buy-back program that started in 2007. The founders of the predecessor companies are still involved in the firm. The current CEO is Tim Brown.

The firm employs approximately 550 people in the disciplines of human factors, mechanical, electrical and software engineering, industrial design, and interaction design. IDEO has worked on thousands of projects for a large number of clients in the consumer, computer, medical, furniture, toy, office and automotive industries. Notable examples are Apple's first mouse, Microsoft's second mouse, the Palm V PDA, and Steelcase's Leap chair. Major clients (as of 2004) included Procter & Gamble, PepsiCo, Microsoft, Eli Lilly, and Steelcase.

In 1999, the firm was the subject of the "Deep Dive" episode of ABC's Nightline; they redesigned a shopping cart in five days. In 2001, IDEO's general manager Tom Kelley wrote *The Art of Innovation*, and more recently, *The Ten Faces of Innovation*.

IDEO was featured in the 2009 design documentary "Objectified".

IDEO has won more of the *BusinessWeek*/IDSA Industrial Design Excellence Awards than any other firm. IDEO has been ranked in the top 25 most innovative companies by businessweek and does consulting work for the other 24 companies in the top 25.

Design Partnership for Samsung

Samsung Electronics has grown to be the most recognized consumer electronics brand in the world. A status earned through a growing commitment to design, Samsung has maintained a close partnership with IDEO since the late 1990s, when the two companies opened a joint office on the IDEO Palo Alto campus. The adoption of design strategy and human-centered design methodologies has helped Samsung differentiate its product line and successfully develop a highly recognized global brand. Some products of Samsung designed by IDEO are shown in Figure 4.39 and Figure 4.40.

Figure 4.39　Keplerphone

Figure 4.40　Detail of 970p monitor

CD Player for Muji

In a 1999 design exploration, then-IDEO designer Naoto Fukasawa conceived this refreshingly simple CD player whose form mimics a ventilation fan as it spins its exposed compact disc. The goal of this project was to identify an essential design approach, and to create simple yet personal, sentient objects. Distilled into its most minimal form, the award-winning CD player consists of a single speaker, no cover, and a vertical power cord (See Figure 4.41).

Figure 4.41　CD Player for Muji

Coasting Bicycle Design Strategy for Shimano

Shimano, the leading supplier of bicycle components, partnered with IDEO in an effort to attract millions of lapsed cyclists and reinvigorate the bike industry. Through fieldwork, the team settled on three program objectives: provide a better riding experience, design an innovative product platform, and enable a more engaging purchasing experience. The resulting strategy, dubbed "Coasting", addressed all three, culminating in a reference design for a bike that has inspired designers, manufacturers, independent dealers, and casual riders alike. (See Figure 4.42—Figure 4.43)

Figure 4.42　Giant Suede Coasting Bike　　　　Figure 4.43　Early Shimano Reference Design

Media: Scape for Steelcase

Steelcase stands at the intersection of workspace systems and technology, connecting people, projects, and teams through leading-edge spaces, experiences, and furniture. Looking for market validation of a first-of-its-kind meeting concept, Steelcase approached IDEO to evaluate the design's potential for success. Using design research, IDEO revealed opportunities to optimize the media-enabled concept and turn it into an award-winning workspace solution.

IDEO and Steelcase set out to understand team dynamics and communication methods, and how team sizes, meeting styles, and available technology can better foster group outcomes. Observations uncovered a desire for simplicity, flexibility, and democracy, which helped guide and inform the design of several evaluative prototypes.

The final design is media: scape, (See Figure 4.44) a "walk-up and connect" experience that merges furniture and technology to help teams access and share information. Designed to look light, open, and inviting, media: scape uses a pedestal table with cleverly embedded hardware and interactive technology to foster group work. A unique puck-shaped handset allows users to connect laptops and control the content of what is projected on a single, or up to four flanking LCD screens. Handsets are easily stowed in the designated "media well", reserving the table surface for active work and related tools. When the system is powered on, first-time users are offered a short tutorial animation, eliminating

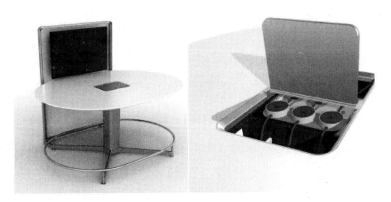

Figure 4.44　Winner of a Gold Medal at NeoCon 2008, Media: Scape Begins Shipping in 2009

the need for IT's help with set-up or shutdown. Available with several material choices, all furniture components are planned for Cradle-to-Cradle certification and include PVC-free cables and wiring.

Key Words

[1] consultancy [kən'sʌltnsɪ] *n.* 1. 顾问(工作) 2. 咨询公司 3. 专家咨询

[2] spinning out 1. 纺…… 2. 尽量使某物延长

[3] episode ['epɪsəud] *n.* 1. (人生的)一段经历；(小说的)片段，插曲 2. (电视连续剧或无线电广播小说的)一集

[4] documentary [dɔkju'mentrɪ] *n.* 纪录片 *adj.* 1. 文件的，文书的 2. 纪实的 3. 纪录的，文献的

[5] electronics [ɪlek'trɔnɪks] *n.* 1. 电子学 2. 电子学应用；电子器件

[6] methodology [meθə'dɔlədʒɪ] *n.* 1. (从事某一活动的)一套方法，原则 2. 方法学，方法论

[7] ventilation [ˌventɪ'leɪʃn] *n.* 1. 空气流通 2. 通风设备；通风方法 3. 公开讨论

[8] reinvigorate [riːɪn'vɪɡəreɪt] *vt.* 使再振作，使复兴

[9] prototype ['prəutətaɪp] *n.* 原型，雏形，蓝本

Key Sentences

1. In a 1999 design exploration, then-IDEO designer Naoto Fukasawa conceived this refreshingly simple CD player whose form mimics a ventilation fan as it spins its exposed compact disc.

在1999年的设计开发中，当时的艾迪欧公司设计师深泽直人构思了这款令人耳目一新的简单CD播放器，这款播放器的形式模仿了通风扇，光盘是旋转且暴露在外的。

2. IDEO and Steelcase set out to understand team dynamics and communication methods, and how team sizes, meeting styles, and available technology can better foster group outcomes. Observations uncovered a desire for simplicity, flexibility, and democracy, which helped guide and inform the design of several evaluative prototypes.

艾迪欧公司和斯蒂尔凯斯着手了解团队动态、沟通方法、团队规模、会议风格和可以促进团队成果的可利用的技术。通过观察，他们发现了简单、灵活和民主的需求，这有助于引导和影响一些评估模型的设计。

课文翻译

IDEO

艾迪欧公司是一家设计和创新咨询公司，坐落在美国加利福尼亚州的帕拉奥托。其他办公室在旧金山、芝加哥、纽约、波士顿、伦敦、慕尼黑和上海。该公司帮助设计产品、服务、环境和数码体验。此外，该公司已越来越多地参与了管理咨询。

艾迪欧公司成立于1991年，由三个已经成立的设计公司合并而成：戴维·克利设计

事务所(由戴维·克利创办,他也是斯坦福大学的教授),ID Two(由英国的比利·莫格里奇创立)和矩阵产品设计事务所(由迈克纳·托尔创立)。办公家具制造商斯蒂尔凯斯拥有该产品多数股份,通过开始于 2007 年的 5 年的产品返销管理程序维持其子公司,公司的创办人仍然参与了该公司。当前的总裁是蒂姆·布朗。

该公司的雇员约 550 人,其学科有人因工程、机械、电气和软件工程,工业设计和交互设计。艾迪欧公司为消费、计算机、医疗、家具、玩具、办公和汽车行业的大量客户设计了数以千计的项目。著名的例子如苹果公司的第一个鼠标、微软的第二个鼠标、Palm V 的掌上电脑以及斯蒂尔凯斯品牌下的"Leap Chair"。2004 年的主要客户包括宝洁、百事可乐、微软、礼来和斯蒂尔凯斯。

1999 年,该公司是美国广播公司(ABC)的"夜线"节目(Nightline)"Deep Dive"的主题,他们在 5 天内重新设计了购物车。2001 年,艾迪欧的总经理汤姆·克利撰写了《发明的艺术》以及最近的《发明的十个方面》。

艾迪欧公司是 2009 年出版的设计纪录片《目标地》(Objectified)的主角。

艾迪欧公司比任何其他公司赢得了更多的《商业周刊》/IDSA 工业设计优秀奖奖项。艾迪欧公司被美国《商业周刊》授予了最具有创造发明力的前 25 个公司之一,并且在这 25 家顶级公司里为其他 24 家公司做设计咨询。

三星的设计伙伴

三星电子已经成长为世界上最具认知性的消费电子品牌。它通过不断增加的设计委托获取了这种地位。自从 20 世纪 90 年代后期,三星就与艾迪欧保持着密切的合作伙伴关系,当时,这两家公司在艾迪欧的帕拉奥托园区联合开设了办公室。设计战略与以人为中心的设计方法的采用帮助三星实现其产品线的差异化,并且成功地发展成为世界上具有高度认知性的全球品牌。

为无印良品设计的 CD 播放器

在 1999 年的设计开发中,艾迪欧的设计师深泽直人构思了这款令人耳目一新的简单 CD 播放器,这款播放器的形式模仿了通风扇,光盘是旋转且暴露在外的。这个项目的目标是确定一种必要的设计方法,并创建简单而个性化的,有感觉的对象。获奖的 CD 播放器被提炼成最小的形式,它包括一个简单的扬声器,没有盖子,还有一根垂直的电线。

希马诺航行车设计战略

希马诺是自行车零件的主要供应商,与艾迪欧合作努力去吸引数以百万的流失的骑车人并且复兴自行车工业。经过调查,这个团队设置了三个程序目标:提供更好的骑车体验,设计一个创新的产品平台,产生更鼓舞人心的购买体验。战略结果被命名为"Coasting,",强调了以上所有三点,以一种参考性设计而告终,它鼓舞了设计师、制造商、独立经销商、漫不经心的骑行者。

斯蒂尔凯斯的"Media:Scape"系列

斯蒂尔凯斯立足于工作空间系统与科技的交叉,通过领先的空间、经验和家具连接人,项目和团队。寻求史无前例的会议概念的市场功效,斯蒂尔凯斯委托艾迪欧评估设计的成功潜力。利用设计研究,艾迪欧揭示了优化媒体概念的机会,并把它变成了一个屡获殊荣的工作区解决方案。

艾迪欧和斯蒂尔凯斯着手了解团队动态、沟通方法、团队规模、会议风格和可以促进团队成果的可用技术。通过观察，他们发现了对简单、灵活和民主的需求，这有助于引导和影响一些评估模型的设计。

最终的设计结果是"Media：Scape"系列，它是一种行走与沟通体验，融合了家具与科技帮助工作组访问和分享信息。设计看起来是明亮的，开放的，有吸引力的。"Media：Scape"使用一张有底座的桌子与嵌入的硬件和交互技术促进团队的工作。一个独特的圆盘状的手写板允许使用者连接笔记本电脑并且控制单个项目的内容，或者放置四台带有侧翼的 LCD 显示屏，手写板被命名为"media well"，易于保存，为积极的工作和相关工具保留桌面。当系统通电，新手用户被提供一段简短的动画教程，从而无需工厂对设置或关机的帮助。提供几种材料的选择，所有的家具部件都计划获得 C2C 认证，包括聚氯乙烯（PVC）自由电缆和电线。

Free Reading 1

Frog Design

Frog design is a global innovation firm. We work with the world's leading companies, helping them create and bring to market meaningful products, services, and experiences. Our cross-disciplinary process reveals valuable consumer and market insights and inspires lasting, humanizing solutions. With a team of more than 500 designers, technologists, strategists, and analysts, we deliver fully convergent experiences that span multiple technologies, platforms, and media. We work across a broad spectrum of industries, including consumer electronics, telecommunications, healthcare, media, education, finance, retail, and fashion. Our clients include Alltel, Disney, GE, HP, Logitech, Microsoft, MTV, Seagate, Yahoo! and others. Founded in 1969, frog is headquartered in San Francisco, California, with studios in Austin, TX; New York, NY; San Jose, CA; Seattle, WA; Milan, Italy; Amsterdam, Netherlands; Stuttgart, Germany; and Shanghai, China.

History

The Early Years

Frog design was founded in 1969 by designer Hartmut Esslinger, around a single guiding principle: "form follows emotion." An adaptation of the familiar phrase "form follows function," the new maxim set the tone for frog's design philosophy by declaring that a product's effect upon its user was as vital as its functionality. Together with partners Andreas Haug and Georg Spreng, Esslinger opened a studio in the Black Forest of Germany to promote this concept of emotional design—viewing every act of creation as a small step towards improving the everyday life of the individual.

Soon after, the company—then Esslinger Design—received its first big break: a commission from German electronics giant Wega. When, a few years later, Wega was bought by Sony, frog found itself working for a massive corporation. The partnership was a huge success, spanning decades and generating more than 100 products, including the mold-breaking black-box Sony Trinitron TV (See Figure 4.45—Figure 4.47). Gradually, the young design firm became known for its innovation, risk-taking, vision, and success.

Figure 4.45　Apple IIc

Figure 4.46　The early partners:
Andreas Haug, Hartmut Esslinger
Georg Spreng

Figure 4.47　TurboChef

Coming to America: Frog Goes Global

It was precisely this mix that appealed to top executive Steve Jobs when he began searching for the elusive magic that would give Apple a market edge. The year was 1981, and computing was a sea of anonymous beige boxes. Jobs combed the world for a strategy-focused design company, and knew he had found it in Esslinger's team. A multimillion-dollar deal was struck, enticing the company to establish a California office. A few years later, the Apple IIC was launched to great fanfare, named "Design of the Year" by TIME Magazine, and inducted into the permanent collection at the Whitney Museum of Art. Apple's revenue soared from US $700 million in 1982 to US $4 billion in 1986.

With the move to California, the company changed its name to frog design—not for its ability to metamorphize, though this would certainly prove the case, but for its international roots: (f)ederal (r)epublic (o)f (g)ermany. The lower case letters offered a nod to the Bauhaus notion of a non-hierarchical language, reinforcing the company's ethos of democratic partnership, both within the design teams and in its client relationships. Now headquartered in San Francisco, frog continues to expand. With its Shanghai studio opening in 2007 and its Amsterdam studio opening in 2008, frog has eight studios globally.

From ID to UI to Convergence

Since then, frog has continuously expanded beyond its industrial design roots, evolving to better address the technological and cultural developments of the marketplace. In the 1980s, the company took on corporate branding, recognizing the value of a consistent user experience across platforms—from product design and engineering to graphics, logos, packaging, and production. Its redesign of the Logitech products and brand identity led the computing giant to raise its revenue from US $43 million in 1988 to over US $200 million in 1995, securing a number-one market position. A few years later, frog reexamined the entire Lufthansa operation—from airport signage and plane interiors to the flatware used for in-flight dining. By exploring the interactions between various consumer touchpoints, frog helped brands create more meaningful product and service experiences.

In the 1990s, frog launched a Digital Media group and began growing its expertise in user interface design for websites, computer software, and mobile devices. Its 1999 redesign of SAP's enterprise software fostered new levels of efficiency in business management worldwide; its 2000 web design of Dell.com set the standard for e-commerce; and its 2001 collaboration with Microsoft helped create the look and feel for Windows XP—touching the lives of millions of consumers across the globe. More recent collaborations in web design, software, mobile devices, and consumer electronics are helping to shape the digital experience still as we move into an era of ubiquitous computing.

In recent years, frog has expanded its offerings once more to include strategic advising on high-level business challenges and long-term planning. Today, the company is a global innovation firm, in which designers, strategists, and technologists work hand in hand to identify the best possible solutions for Fortune 500 clients. With nine studios internationally and more than 400 employees, frog serves many of today's top businesses, including Alltel, Disney, GE, HP, Logitech, Microsoft, MTV, Seagate, Siemens, Yahoo! and others.

Disney Consumer Electronics

Challenge

Disney, the world's foremost storytellers, wanted to bring serious consumer electronics to its target demographic: kids. Industry observers had already dismissed this market segment as a dead end but given its strong brand recognition, emotional resonance, and unique customer loyalty, Disney recognized it was uniquely suited to address this market. The challenge was to bring new ideas to life and infuse an entire line of products with Disney magic. The project was immense in scope: end-to-end product development and distribution, from initial concept and brand strategy to product packaging and multimedia retail distribution.

The Depth of Disney

When this project began, there were no consumer electronics for kids. "My First Sony," an unsuccessful attempt to break into this market 10 years earlier, provided valuable lessons: it was too expensive, lacked breadth, and looked too much like a toy.

Disney is a content company. To make its offerings unique, it was important to leverage its brand identity, integrate content elements, and expand the world of its beloved characters through the products—but not in superficial ways such as inserting DVDs, or scaling back the functionality of the products. In this way, Disney could create products that emotionally resonate with kids, and have features that get parents to buy the product.

Delving deeper into the "wonderful world of Disney" revealed things that most people never consciously notice, such as the perfect position of Mickey Mouse's ears—they never move, regardless of the position of his head, to deliver a consistent silhouette. There are many common elements of Disney characters, such as a pervasive sense of asymmetry that gives them a feel of constant movement, creating a sense of urgency and excitement.

Practical Magic

These details informed the early conceptual work around the new products. The goal wasn't to create a "Mickey Mouse phone", but a phone that lives in the same universe that created Mickey Mouse (See Figure 4.48).

Establishing an exciting visual design language laid the foundation for the entire product family. The products feature unique graphic user interfaces and sounds that blur the lines between the device, the interface, and the Disney content. Compelling graphic design and packaging, as well as in-store displays featuring animated video loops with music, capture the magical Disney content and bring these products to life in the retail environment.

Figure 4.48 Bold colors and soft shapes perfectly capture the magic of Disney, positioning it as a powerful leader for children's consumer electronics.

Bringing Businesses Together

Disney started out with the "bare necessities" of an idea for a line of products. After formulating the product development and distribution strategy, it was time to meet with manufacturers, retailers and select partners to make the idea a reality. Motorola partnered with retailers in the actual conceptual design process for the Disney cordless phone and the FRS 2-way radio, and worked closely with frog and Disney every step of the way until the products appeared on store shelves. The close nature of this work enabled engineers to talk with manufacturers every step of the way and show how innovative concepts could be realized at a low cost.

Result

These whimsical products were built for children to love, but the process that created them has an equally important business impact. The products are distributed worldwide and will yield US $500 million in sales this year. This partnership empowered Disney to leverage its brand in a whole new, more meaningful way. Rather than passively deal with licensing its content to third-party design and manufacturing companies, Disney now drives every aspect of its new products. They control what the products look like and how they act in the marketplace. As a result, Disney expanded the reach of its brand, its resonance with consumers, and its role as a product company delivering consumer electronics.

Lufthansa Brand Experience

Challenge

Lufthansa initially approached frog to streamline the check-in process at its Frankfurt Airport terminal, aware of the great impact this interaction has on the brand experience. But after discussions with the frog team, the airline decided to explore a more integrated, holistic design approach. Together, frog and Lufthansa set about revitalizing the entire brand experience across operations, services, and environments, imbuing the airline with an aura of technological sophistication and old-fashioned romance.

Process

Customer Delight

We conceived of each trip on Lufthansa as a continuous experience, stretching from the initial human interaction of check-in, through the terminal, lounge, and gates, and into the airplane itself. It wasn't enough to redesign one part of this experience alone; every element needed to underscore the company's existing equity—its reputation for reliability—with a new emphasis on friendliness, comfort, and relaxation. Together, frog and Lufthansa examined every aspect of the company's identity: its brand architecture; the use of colors in its logo; the primary and secondary core values. We studied the experiences of passengers and staff to identify where improvements could be made. We sought to renew the importance of the airport as a place and as a space.

Retro-Futurism

Our research indicated that the design of departure lounges actually adds to the stress of flying. Consequently, frog pursued an agenda of environmental design that would cast maximum convenience as its central ethos. sweeping aerodynamic curves and non-traditional seating patterns in the lounge area smooth away passengers's anxiety, showers, conference areas, cafes, and configurable furniture create a comfortable and enjoyable travel experience that tailors to the personal needs of its visitors.

By blending nostalgia for the rich, romantic history of aviation with the professional, technical expertise of the brand today, frog's designers produced a unique visual lexicon for Lufthansa. The shapes and materials of the interiors borrow from the language of airplanes, with the curved forms and metallic finish of a wing, the crisp professionalism of a pilot's uniform. The graphics team developed a distinct, highly legible signage system that further communicates this singular marriage of efficiency and warmth.

We then extended this language into the planes themselves, designing seats so supportive and comfortable that customers can shed their anxieties and actually enjoy the experience of flying. First-class seats convert to beds with storage space for personal belongings. Business-class seats offer a personal video screen and—in a then-unprecedented gesture of stylishness and functionality—headrests whose side flaps could be adjusted for extra support. Our attention to detail included even the redesign of footrests, water-bottle holders, and in-flight flatware.

The Mechanics of Change

In collaboration with architect Michael McDonough, enhancements were made to the entire Lufthansa terminal area at Frankfurt Airport. Deploying technical engineers, model makers, and designers, frog produced the full mechanical design specifications for the redesign of the gate areas and departure lounges. A journey on Lufthansa, blending glamour and luxury, is now a pleasurable adventure from beginning to end.

25-year-old Porsche Design

It was in 1972 that F. A. Porsche left his father Ferry's automobile company and set up his own design firm in Stuttgart. He moved to Zell-am-See two years later and continues to breathe the air of these beautiful natural surroundings today. From there, Porsche Design is sending messages to the world.

Porsche Design has a staff of 12 persons: President Porsche, Managing and Design Director Dirk Schmauser (who is known for tough negotiating and superior management skills), six designers, two modelers and two secretaries. Though personnel changes tend to be frequent in the design world, surprisingly at Porsche Design the staff has remained almost unchanged since 1979, and has always consisted of about ten people. Last year a computer expert was hired in response to the recent digital designing trend.

Almost all the members of the staff had experience designing in-house for automobile or home appliance manufacturers or as free-lance designers before joining Porsche Design. They also serve as guest lecturers at universities or specialized colleges with design courses. Though small, this is truly a top class group of individuals.

The staff members live within 10 minutes from work place, so they do not have to put up with the time, effort and stress of commuting. On our visit the six designers did not seem overly busy, but they are always working on three or four projects at once. They are in healthy competition with each other until the person responsible for development is determined and presentations to the client begin. The designers respect each other, and one of the staff members told us that the head management are excellent directors for the designers who are immersed in their development work.

History and Works

F. A. Porsche is a direct grandson of Prof. Dr. Ferdinand Porsche and the eldest son of Ferry Porsche, both creators of the renowned 356 Series. He began naturally working at the sports car company at the age of 23 and provided brilliant contributions. He was appointed Chief Designer in 1962. He developed F1 cars and the 904 Carrera, and in 1963 created his masterpiece, the 911. He was only 28 at the time.

In 1968, F. A. Porsche became a member of the Porsche KG. but withdrew in 1972 and established Porsche Design along with a staff of four. That very year the company developed a chronograph wristwatch. The next year it ventured into the field of optical equipment, working on a 35 mm single-lens reflex camera, small binoculars and an 8 mm cinematic camera. In 1974 the Karl Zeiss Foundation with the cooperation of Yashica mass-produced the Contax camera designed by Porsche Design.

The firm moved to Zell-am-See in 1974 and the staff grew to six. At first it worked on products for children—not amazingly since he had two little boys running around at home. These included a safety seat, a cart that also transforms into a tricycle, and the "Jagowagen" off-roader. Though many did not get past the prototype stage, these were full-fledged projects with an emphasis on safety. In 1977 Vespa mass-produced an engined cart designed by Porsche Design.

Between 1976 and 1980 Porsche worked on a number of transportation projects, including the Steyr-Puch Cobra GT80/6 (a lightweight motorcycle) and a small jet revealed to the public at the Atlanta Air Show. Both were mass-produced. This was also the time of the "Motorcycle of Tomorrow" project held by the German motorcycle magazine "Motorrad". This was a competition between Porsche Design, Hans Muth and Giorgetto Giugiaro. F. A. Porsche announced a prototype built with the cooperation of Yamaha.

The Birth of the Porsche Design Brand

There were two design development projects that made this design house world famous. The first was measuring equipment for BBC, a Swiss manufacturer of heavy electrical machinery. Their design broke with all accepted projects and received high praise, including an award from the Austrian government.

The second was the series of merchandise developed by Porsche Design Products. These products carrying the Porsche Design brand had designs stressing function and were thoroughly permeated with F. A. Porsche's design philosophy. Porsche Design established partnerships with IWC for watches and Carrera for goggles and sunglasses and began developing products with revolutionary ideas. Porsche Design Products was set up in Salzburg in 1979. The company's name was later changed to Porsche Design Management (PDM). It involved only in license management. F. A. Porsche's eldest son Oliver serves as its Managing Director. Porsche Design products have now established a firm market and known by people throughout the world for their extremely high quality.

In 1980 the staff of Porsche Design grew to eight, and Dirk Schmauser, who had previously worked at Siemens and Audi, was taken on as Managing and Design Director.

In 1981 Porsche Design worked on its first project with a Japanese company, a telephone enveloped in soft synthetic rubber for NEC. The Yamaha compact headphones Porsche Design developed in 1984 are on permanent display at the Museum of Modern Art in New York.

In the 1980s Porsche Design developed many pieces of award winning interior decor and lighting fixtures for Italian companies, including the Artemide "Mikado" lighting system and the Italiana Luce "Jazz".

Below you will find the major works of Porsche Design right up to the most recent ones. When viewing them, you will no doubt understand why this company has for so long stuck to its precept of "Libertise and Limits". The design trends for many Japanese products, including automobiles, are now clearly changing. The concise forms and ergonomic designs of these new Japanese projects finally seem to be tuning to the same wavelength as the refined forms stressing function that Porsche Design has been creating for the past 25 years. Some works of Porsche Design are shown in Figure 4.49—Figure 4.52.

Figure 4.49 Karl zeiss/yashica "contax RTS" (1974) The world's first electronically controlled camera, designed by F. A Porsche With manufacturing technology provided by Yashica.

Figure 4.50 "2001" telephone for the German Federal Postal Administration (1989)

Figure 4.51 Italiana Luce "JAZZ" table lamp (Italy, 1989)

Figure 4.52 Barazzoni "LUCIE OMBRE" pot and pan system (Italy, 1989) Pots and pans with aesthetic that take account of the meal as a social ritual and allowing use on both the stove and the table. The lid can be used as a stand to protect the table. The ripples on the lid cause the steam to return equallto the top of the food, and the handle on the side allows easier stacking. 1989 Compasso d' Ore design prize (Milan).

Free Reading 3

Museum of Modern Art

The Museum of Modern Art (stylized MoMA) is an art museum located in Midtown Manhattan in New York City, on 53rd Street, between Fifth and Sixth Avenues (See Figure 4.53). It has been singularly important in developing and collecting modernist art, and

Figure 4.53 The entrance to The Museum of Modern Art

is often identified as the most influential museum of modern art in the world. The museum's collection offers an unparalleled overview in modern and contemporary art, including works of architecture and design, drawings, painting, sculpture, photography, prints, illustrated books and artist's books, film, and electronic media.

MoMA's library and archives hold over 300,000 books, artist books, and periodicals, as well as individual files on more than 70,000 artists. The archives contain primary source material related to the history of modern and contemporary art. It also houses an award-winning fine dining restaurant, the Modern, run by Alsace-born chef Gabriel Kreuther.

History

The idea for the Museum of Modern Art was developed in 1928 primarily by Abby Aldrich Rockefeller (wife of John D. Rockefeller Jr.) and two of her friends, Lillie P. Bliss and Mary Quinn Sullivan. They became known variously as *"the Ladies"*, *"the daring ladies"* and *"the adamantine ladies"*. They rented modest quarters for the new museum in rented spaces in the Heckscher Building at 730 Fifth Avenue (corner of Fifth Avenue and 57th Street) in Manhattan, and it opened to the public on November 7, 1929, nine days after the Wall Street Crash. Abby had invited A. Conger Goodyear, the former president of the board of trustees of the Albright Art Gallery in Buffalo, New York, to become president of the new museum. Abby became treasurer. At the time, it was America's premier museum devoted exclusively to modern art, and the first of its kind in Manhattan to exhibit European modernism.

Goodyear enlisted Paul J. Sachs and Frank Crowninshield to join him as founding trustees. Sachs, the associate director and curator of prints and drawings at the Fogg Art Museum at Harvard University, was referred to in those days as a collector of curators. Goodyear asked him to recommend a director and Sachs suggested Alfred H. Barr Jr., a promising young protégé. Under Barr's guidance, the museum's holdings quickly expanded from an initial gift of eight prints and one drawing. Its first successful loan exhibition was in November 1929, displaying paintings by Van Gogh, Gauguin, Cézanne, and Seurat.

First housed in six rooms of galleries and offices on the twelfth floor of Manhattan's Heckscher Building, on the corner of Fifth Avenue and 57th Street, the museum moved into three more temporary locations within the next ten years. Abby's husband was adamantly opposed to the museum (as well as to modern art itself) and refused to release funds for the venture, which had to be obtained from other sources and resulted in the frequent shifts of location. Nevertheless, he eventually donated the land for the current site of the museum, plus other gifts over time, and thus became in effect one of its greatest benefactors.

During that time it initiated many more exhibitions of noted artists, such as the lone Vincent van Gogh exhibition on November 4, 1935. Containing an unprecedented sixty-six oils and fifty drawings from the Netherlands, and poignant excerpts from the artist's letters, it was a major public success and became "a precursor to the hold van Gogh has to this day on the contemporary imagination".

The museum also gained international prominence with the hugely successful and now famous Picasso retrospective of 1939—1940, held in conjunction with the Art Institute of Chicago. In its range of presented works, it represented a significant reinterpretation of Picasso for future art scholars and historians. This was wholly masterminded by Barr, a Picasso enthusiast, and the exhibition lionized Picasso as the greatest artist of the time, setting the model for all the museum's retrospectives that were to follow.

When Abby Rockefeller's son Nelson was selected by the board of trustees to become its flamboyant president in 1939, at the age of thirty, he became the prime instigator and funder of its publicity, acquisitions and subsequent expansion into new headquarters on 53rd Street. His brother, David Rockefeller, also joined the museum's board of trustees, in 1948, and took over the presidency when Nelson took up position as Governor of New York in 1958.

David subsequently employed the noted architect Philip Johnson to redesign the museum garden and name it in honor of his mother, the Abby Aldrich Rockefeller Sculpture Garden. He and the Rockefeller family in general have retained a close association with the museum throughout its history, with the Rockefeller Brothers Fund funding the institution since 1947. Both David Rockefeller, Jr. and Sharon Percy Rockefeller (wife of Senator Jay Rockefeller) currently sit on the board of trustees. In 1937, MoMA had shifted to offices and basement galleries in the Time & Life Building in Rockefeller Center. Its permanent and current home, now renovated, designed in the International Style by the modernist architects Philip Goodwin and Edward Durell Stone, opened to the public on May 10, 1939, attended by an illustrious company of 6,000 people, and with an opening address via radio from the White House by President Franklin D. Roosevelt. *Wish Tree*, Yoko Ono's installation in the Sculpture Garden (since July 2010), has become very popular with contributions from all over the world. In 1997 the Japanese architect Yoshio Taniguchi beat out ten other international architects to win the competition to execute the redesign of the museum, which after being closed in Manhattan for a time during the process (a temporary space was opened in Long Island City, Queens) reopened in 2004.

Artworks

Considered by many to have the best collection of modern Western masterpieces in the world, MoMA's holdings include more than 150,000 individual pieces in addition to approximately 22,000 films and 4 million film stills. The collection houses such important

and familiar works as the following:
- The Starry Night by Vincent van Gogh(See Figure 4.54)
- The Sleeping Gypsy by Henri Rousseau (See Figure 4.55)

Figure 4.54 Vincent van Gogh, The Starry Night

Figure 4.55 Henri Matisse, The Dance I, 1909

- The Dream by Henri Rousseau
- Les Demoiselles d'Avignon by Pablo Picasso (See Figure 4.56)
- The Persistence of Memory by Salvador Dalí(See Figure 4.57)

Figure 4.56 Pablo Picasso, Les Demoiselles d'Avignon, 1907

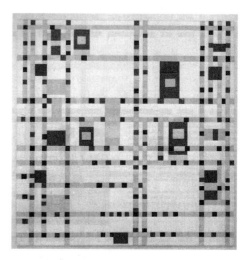

Figure 4.57 Piet Mondrian, Broadway Boogie Woogie, 1942—1943

- Broadway Boogie Woogie by Piet Mondrian
- Campbell's Soup Cans by Andy Warhol
- Te aa no areois (The Seed of the Areoi) by Paul Gauguin
- Water Lilies triptych by Claude Monet
- The Dance by Henri Matisse
- The Bather by Paul Cézanne
- The City Rises by Umberto Boccioni
- Love Song by Giorgio De Chirico
- Number 31, 1950 by Jackson Pollock

- Vir Heroicus Sublimis by Barnett Newman
- Broken Obelisk by Barnett Newman
- Flag by Jasper Johns
- Christina's World by Andrew Wyeth
- Self-Portrait With Cropped Hair by Frida Kahlo
- Painting (1946) by Francis Bacon
- Two Children Are Threatened by a Nightingale by Max Ernst

It also holds works by a wide range of influential European and American artists including Georges Braque, Marcel Duchamp, Walker Evans, Helen Frankenthaler, Alberto Giacometti, Arshile Gorky, Hans Hofmann, Edward Hopper, Paul Klee, Franz Kline, Willem de Kooning, Dorothea Lange, Fernand Léger, Roy Lichtenstein, Morris Louis, René Magritte, Aristide Maillol, Joan Miró, Henry Moore, Kenneth Noland, Georgia O'Keeffe, Jackson Pollock, Robert Rauschenberg, Auguste Rodin, Mark Rothko, David Smith, Frank Stella, and hundreds of others.

MoMA developed a world-renowned art photography collection, first under Edward Steichen and then John Szarkowski, which included photos by Todd Webb, as well as an important film collection under the Museum of Modern Art Department of Film and Video. The film collection owns prints of many familiar feature-length movies, including *Citizen Kane and Vertigo*, but the department's holdings also contain many less-traditional pieces, including Andy Warhol's eight-hour *Empire* and Chris Cunningham's music video for Björk's *All Is Full of Love*. MoMA also has an important design collection, which includes works from such legendary designers as Paul László, the Eameses, Isamu Noguchi, and George Nelson. The design collection also contains many industrial and manufactured pieces, ranging from a self-aligning ball bearing to an entire Bell 47D1 helicopter.

Renovation

MoMA's midtown location underwent extensive renovations in the early 2000s, closing on May 21, 2002 and reopening to the public in a building redesigned by the Japanese architect Yoshio Taniguchi, on November 20, 2004. From June 29, 2002 until September 27, 2004, a portion of its collection was on display in what was dubbed *MoMA QNS*, a former Swingline staple factory in Long Island City, Queens.

The renovation project nearly doubled the space for MoMA's exhibitions and programs and features 630,000 square feet (59,000 m^2) of new and redesigned space. The *Peggy and David Rockefeller Building* on the western portion of the site houses the main exhibition galleries, and *The Lewis B. and Dorothy Cullman Education and Research Building* on the eastern portion provides over five times more space for classrooms, auditoriums, teacher training workshops, and the museum's expanded Library and Archives. These two buildings frame the enlarged *Abby Aldrich Rockefeller Sculpture Garden*.

MoMA's reopening brought controversy as its admission cost increased from US $12 to US＄20, making it one of the most expensive museums in the city; however it has free entry on Fridays after 4pm, thanks to sponsorship from Target Stores. Also, all CUNY college students receive free admission to the museum by simply going to the information desk. The architecture of the renovation is controversial. At its opening, some critics thought that Taniguchi's design was a fine example of contemporary architecture, while many others were extremely displeased with certain aspects of the design, such as the flow of the space.

MoMA has seen its average number of visitors rise to 2.5 million from about 1.5 million a year before its new granite and glass renovation. The museum's director, Glenn D. Lowry, expects average visitor numbers eventually to settle in at around 2.1 million.

Core Text 14

Responsible Products: Selecting Design and Materials

Sustainable Development and Methods for Evaluating it

For all intents and purposes, responsible design is synonymous with sustainable development. Both consider the following issues:

- Environment. The world's ecosystem must be respected. The environment must be seen as the theater for any product development process. It must not be considered to be free to use and abuse.
- Equity. The issues of poverty, equal distribution, and use of natural resources must be solved in order to maintain stable and egalitarian societies.
- Economy. A reasonable profit and the feasibility of selling the product on the market are, of course, also important issues.

One way to create a systematic environmental plan for product development is by using an environmental management system (EMS), such as ISO 14000 (which is an international environmental standard) or the European regulation, EMAS (European Eco Management and Audit Scheme). These have become increasingly prevalent worldwide. They are, of course, voluntary systems; in fact, any company that has taken steps toward addressing environmental issues can be said to have implemented an EMS.

The Life Cycle Assessment (LCA) methodology is probably the best-known technique for evaluating environmental impacts associated with a certain product. LCA is an example of a standardized "tool" in the ISO 14000 standard and is one important part of creating an EMS. ISO 14000 also includes standards for labeling products, performing audits, and evaluating environmental performance. LCA evaluates a product step by step, "from cradle to grave". There are typically six steps in an LCA evaluation:

1. Extract from material
2. Manufacturing and refinement
3. Packaging
4. Transportation of material and finished product
5. Product use
6. Product disposal

Bear in mind that responsible design alone is not sufficient to sell a product. If there isn't a market for it, it doesn't matter how sustainable its development is. Taking the temperature of a market requires close contacts with customers. For instance, cultural aspects are critical. Take flooring, for example. In many countries, wooden floors are popular because of their appearance and relative softness. Concrete is both more durable and more sustainable, but concrete flooring is not popular for many customers. It is important to note that such cultural differences are also a matter of the availability of materials. In highly populated countries, wood, for example, is typically uncommon and expensive. In such countries, stone, concrete, and steel are more often used as structural building materials. In the wood-rich countries of Northern Europe, on the other hand, wood is actually a cheap alternative to other materials.

Selection of Materials

Structural materials can be divided into six groups: metals, ceramics, synthetic polymers, and natural organic, inorganic, and composite materials. There are pros and cons to each.

Metals are typically cheap and easy to recycle into new products. On the other hand, the high weights/densities of many metals (steel and copper, for instance) mean much energy is consumed in transporting them. Recycling metals also requires high expenditures of energy due to typically high melting points. Steel, especially, is not very resistant to corrosion, and protecting it often requires toxic chemical treatments or paints. Aluminum, on the other hand, is quite corrosion-resistant, as well as lightweight and strong.

Ceramics are typically lightweight and nontoxic. They are also quite durable, although they can often be brittle. Because they must be crushed, ground, and reburnt, which is typically more costly and requires more energy than producing them from new material, and because the raw materials needed to make them are fairly inexpensive, ceramics are seldom recycled into new products. On the other hand ceramics can, because of their environmentally friendly properties, be crushed and used as fillers in building projects. Forming ceramics by sintering (burning) consumes a fair amount of energy.

Synthetic polymers or plastics: Plastics are often seen as inferior materials. They have frequently been used as a cheap substitute for more-expensive materials, and they have also been used to create poorly made products. However, if materials are selected carefully, polymers can be a good alternative, even the best choice, for many parts or products.

Plastics are generally made from raw oils whose source is limited, and they are therefore considered nonrenewable materials. Once these oils are made into polymers, they can't be recovered to oil again in an easy way. Energy recycling of plastics can be done through burning. The carbon dioxide formed in the process is harmless in small amounts, but it does increase the atmospheric greenhouse effect. Burning is suitable for simple plastics like polyethylene (PE), and such plastics can even be mixed with paper waste and energy recycled in a quite environmentally friendly way to get energy for heating, for instance, in thermal power stations.

However, there are many polymers, especially those containing halogens, like Teflon (PTFE) and polyvinyl chloride (PVC), which release harmful fumes when burned. These are better recycled by remelting, and they can be formed into new products, although controlled burning (in which the exhaust is cleaned) is an alternative. Recycling plastics is sometimes problematic because of the need for determination of the polymer type. However, international labeling has minimized this problem. Natural materials like wood and cotton are quite popular. Recycling these materials is typically a matter of recycling energy. When they are burned, they can, for instance, heat our homes. This process also produces carbon dioxide and water; however, in this case the carbon dioxide can be taken directly from the air by plants and trees. This means that the carbon dioxide content in our atmosphere will not increase through the burning of natural organic materials, and so these materials are often called renewable. Wood does work well for sustainable products. However, it breaks down easily, and painting and impregnating it with chemicals to slow this process can make it toxic, as the toxicants are released to the air or groundwater as a result of burning or putrefaction.

Natural inorganic materials, such as stone, exhibit much the same characteristics as ceramic materials; they are just not synthetic, as ceramics are. Natural stone, for instance, is a natural material with ceramic properties. Unlike ceramics, no extra energy is required to create it, since it already exists in a solid form naturally. However, if forming is needed, natural stones are expensive to form because of their material characteristics.

Composite materials are sometimes difficult to recycle, depending on the materials from which they are made, which sometimes need specific separation. The limited use of material in general for composites and their typical low weight is favorable for vehicles, for example, because it cuts down on fuel consumption. The energy used for production is typically comparable with the energy used for the specific material components in the composite. The mix of such materials as, for instance, glass fibers in polymers is time-consuming for large items like boats, but it seldom leads to extra energy consumption during manufacturing. Some practice case of material application are shown in Figure 4.58—Figure 4.61.

Figure 4.58 This building features on outer panel mode of a durable ceramic (brick), stone plates on the roof, and sheet-copper-plated ports. The building, Sofia Church in Jonkoping, Sweden, was built during the late 1800s. The brick is likely to lost for many hundred years; the stone plates and copper sheeting will probably be somewhat less durable. This is on example of design for low-intensity service, as well as for on extended life through selection of building materials.

Figure 4.59 The liquid medical suspension on the left side is designed with a double package consisting of a plastic container, as well as an outer proper package. In contrast, the toothpaste package at right is designed to minimize waste.

Figure 4.60 The number 02 recycling mark on this synthetic polymer toy car indicates that the product is mode of polyethylene. Such standardized labeling makes it easier to sort the product into the correct group and to recycle it.

Figure 4.61 The compact fluorescent lamp on the left typically gives about five times more light for the some energy consumption as the traditional filament version (in other words, a 10-watt fluorescent lamp would give the some intensity of light as a 50-watt filament lamp). This is a typical example of increased efficiency during the usage phase. Because fluorescent bulbs tend to be longer than the filament versions, it is important to design the reflectors for the electric fittings accordingly.

Final Recycling and Disposal

The last steps in an LCA process deal with recycling and final utilization of material. However, even in the early stages of product development and design, recycling must be taken into account if one is to manufacture a sustainable product with minimum of environmental impact.

Recycling materials in a product that has served its time can be done in several ways:

- *Energy recycling*. The material is burned and the energy is used to create heat.
- *Material recycling*. Material is typically re-melted or reformed in order to make new products.
- *Reuse (or recycling)*. The product is used again, sometimes after reconditioning, as spare parts sold at lower prices than new ones. This is typical for secondhand markets (used car parts, and so forth).
- *Breakdown*. Natural materials, plastics, and some other materials can be broken down naturally or by chemical methods into products or chemicals that are environmentally friendly. These are often referred to as degradable or biodegradable materials and they are often broken down through composting or decomposition.
- *Deposition*. If a product is not suitable for the foregoing processes, the material

can be deposited. However, deposition is often restricted by environmental legislation.

Note that the word recycling can be interpreted in several ways, as seen in the examples above. It's a good idea to offer a clear definition of the word to avoid misunderstandings.

Some general advice for more-efficient use of materials and minimized environmental impact:

- Minimize materials used directly or indirectly for energy, as well as any toxic substances related to the product during its lifecycle.
- Substitute materials to reduce the environmental impact.
- Practice material-efficient design. This means using as little material as possible. To obtain high strength in structural products, the Finite Element Method (FEM) and similar systems can be used to optimize the strength of the product. Use only thin layers of rare or expensive materials—for example, wood veneers or plating of noble metals on cheaper substrates.
- A substitution of composite materials can sometimes reduce the total amount of material used in a construction because of the often better utilization of the materials used. However, it is generally more problematic to recycle mixed materials; this must be considered against the production benefits.
- Materials selection and product design should always take into account future recycling of the materials used.
- Use light materials when manufacturing vehicles, for instance, to reduce energy consumption.

Material changes in a product are generally meant to improve its performance. Every material has its own properties such as strength or low weight which must be adapted to the product. The design is therefore closely related to the materials used.

Key Words

[1] sustainable [səˈsteɪnəbl] *adj*. 1.（对自然资源和能源的利用）不破坏生态平衡的，合理利用的 2. 可持续的

[2] egalitarian [ɪˌgælɪˈteərɪən] *adj*. 主张平等的，平等主义的 *n*. 平等主义

[3] systematic [ˌsɪstəˈmætɪk] *adj*. 成体系的；系统的；有条理的；有计划，有步骤的

[4] synthetic polymers 人造聚合物；合成高分子材料

[5] ceramics [səˈræmɪks] *n*. 1. 制陶术，陶器制造；陶瓷工艺 2. 陶器，陶瓷

[6] toxic [ˈtɒksɪk] *adj*. 有毒的；因中毒引起的

[7] sintering [ˈsɪntərɪŋ] *v*. 烧结

[8] polyethylene [ˌpɒlɪˈeθɪliːn] *n*. 聚乙烯

[9] Teflon *n*. 特氟纶，聚四氟乙烯（商标名称）

[10] chloride [ˈklɔːraɪd] *n*. 氯化物

[11] inorganic [ɪnɔː'gænɪk] adj. 1. 无机的 2. 无组织结构的 3. 体外的
[12] disposal [dɪ'spəʊzəl] n. 清除，处理，处置 adj. 处理（或置放）废品的
[13] substitution [sʌbstɪ'tjuːʃən] n. 代替；代用；替换
[14] strength [streŋθ] n. 1. 力，力量；体力，力气 2. 强度；浓度

课文翻译

负责任的产品：选择设计和材料

可持续性发展和评估方法

从每一层意义来说，负责任的设计等同于可持续发展。二者需要考虑以下问题：

- 环境。世界生态系统必须受到尊敬。环境被认为是任何产品发展过程的剧场。它不应被自由地使用和滥用。
- 权益。为了保持稳定和平等的社会。贫困问题、平等分配和自然资源的使用必须得到解决。
- 经济。当然，合理的利润和市场上销售产品的可行性也是重要的问题。

为产品开发创造系统环境计划的一个方法是使用环境管理体系（EMS），如 ISO14000（这是国际环境标准）或欧洲的规定，EMAS（欧洲 Eco 管理和审计计划）。这些已经广泛流行。当然，它们是自愿的系统；事实上，任何公司已采取步骤处理环境问题，可以说已经执行了环境管理系统。

生命周期评估（LCA）方法可能是最著名的评估某种产品对环境影响的技术。LCA 是 ISO14000 标准系统中标准化的"工具"的实例，是创建环境管理体系的一个重要部分。ISO14000 也包括标准的产品标签、审计和评估环境的性能。LCA 逐步评估产品，"从摇篮到坟墓"。这是典型的 LCA 评估的六个阶段：

1. 从材料中提取
2. 制造和精炼
3. 包装
4. 材料和最终产品的运输
5. 产品使用
6. 产品处理

要记住，只负责设计对推销产品是不够的。如果没有市场，无论多么可持续的发展也无所谓。应将市场需求的温度与消费者密切联系。例如，文化方面是至关重要的。我们以地板为例。在许多国家，木地板受欢迎，因为它们的外表相对柔和。混凝土更长久，更可持续，但混凝土地板对许多客户来说不受欢迎。值得注意的是，这种文化的差异也是材料的可利用性问题。在人口稠密的国家，木材通常是稀罕和昂贵的。在这些国家，石头、混凝土、钢材是常用结构建筑材料。另外，在北欧盛产木材的国家，木材与其他材料相比是相对便宜的。

材料的选择

结构性材料被划分为 6 组：钢材、陶瓷、合成高分子材料、天然有机材料、无机材料、合成材料，每一种都有利有弊。

金属一般都是廉价的，容易回收并产生新产品。另外，许多金属高重量或者高密度

(例如钢铁、铜)意味着在运输它们的过程要消耗更多的能量。回收金属由于高熔点也需要高昂的能源代价,尤其是钢材不是很耐腐蚀,这常常需要有毒化学处理或喷漆。另外,铝耐腐蚀,同时质量轻而且坚固。

陶瓷具有轻和无毒的典型特征,也相当耐用,虽然它们很容易碎。因为它们必须被压碎,粉碎,重燃烧,这个过程比通过新材料重新生产更昂贵并且需要更多的能源,由于原材料使它们相对比较便宜,陶瓷很少循环再生成新产品。另一方面,陶瓷因为其环保属性,在工程项目中压碎被用作填充物。通过烧结(燃烧)形成陶瓷要消耗大量能源。

合成聚合物或者塑料:塑料经常被认为是劣质材料。它们经常被用作更昂贵材料的廉价代用品,也被用来制作拙劣的人造产品。然而,如果仔细选择材料,聚合物是很好的选择。对于许多部件或者产品来说,甚至是最好的选择。

塑料通常是由资源有限的原油制成,因此被认为是不可再生材料。一旦这些油被制成聚合物,它们不会以一种简单的方式重新变成油。塑料的能量循环可以通过燃烧进行。该过程产生的二氧化碳有少量危害,并提高大气中的温室效应。燃烧适合简单的塑料,如聚乙烯(PE),这种塑料甚至可以与造纸废品混合,并且以一种环保的方式进行能量循环为加热获得能源,例如,热发电站。

然而,许多聚合物,尤其是那些含有卤化物的,像聚四氟乙烯(PTFE)和聚氯乙烯(PVC),在燃烧时释放有害的烟雾。通过重熔可以更好地回收,并且可以形成新产品,虽然控制燃烧(在这个过程中排气是清洁的)是可选择的。回收塑料有时是个问题,因为这要取决于聚合物的类型。然而,国际化标记已经减轻这个问题所带来的危害。天然原料,像木材、棉花很流行。回收利用这些材料是典型的回收能源。如果将它们燃烧,可以立刻加热我们的房间。这个过程也产生二氧化碳和水。然而,在这种情况下,二氧化碳可以通过植物和树木直接从空气中提取。这意味着在我们的大气中,二氧化碳的含量通过燃烧自然有机物质不会增加,所以这些材料经常被称为再生的。木头适用于可持续的产品。然而,它容易损坏,通过喷漆和填充化学药品减慢了这一过程,使它产生毒性,当燃烧或者腐烂时,会对空气和地下水释放有毒物质。

自然非有机材料,如石头,展现了与陶瓷相同的属性,它们像陶瓷一样是非人造的。如自然的石材是与陶瓷的属性类似的自然材料。与陶瓷品不同,在创建它时不需要额外的能量,因为它已经以固体状态存在。然而,由于其材料特性,天然石材形成是很昂贵的。

复合材料通常很难回收,根据组合成它们的材料类型,它们有时需要特别分离。有限使用的材料多数为复合材料,其典型的低重量适合交通工具,例如,它降低了燃料消耗。在复合材料中用来生产的能源和用于特定材料组成部分的能源是相当的。这种材料,例如聚合物中的玻璃纤维的混合物对于诸如船的大型物品是耗时的,但在制造过程中很少会导致额外的能量消耗。

最后的循环和处理

生命周期评估过程的最后一步是处理循环和材料的最后利用。然而,即使在产品设计开发的早期阶段,如果制造一种可持续性的产品使其对环境的影响减至最小,必须考虑到循环再利用

产品回收材料可以通过以下几个途径节约时间:
- 能量循环。材料被燃烧,其能源用于产生热量。
- 材料循环。材料被重新融解或者革新以产生新的产品。

- 再利用(或回收)。该产品还可以再用的,有时在条件改变后,部分零件价格比新的要便宜,这是典型的二手市场。
- 分解。天然原料、塑料和一些其他材料可自然分解或通过化学方法转为对环境友好的产品或化学品。这些常被称为降解或可自行分解的物质,它们经常通过堆制肥料或者分解衰减。
- 降解。如果产品不适合上述过程,材料可以被降解。然而,降解通常受严格的环境立法限制。

应注意的是,回收这个词可以解释为几个方面,如上面的例子。给这个词提供明确的定义,避免误解是一个好主意。

一些更有效的使用材料和减少对环境影响的建议:
- 减少直接或者间接耗费能源的材料,以及任何在产品生命周期内的有毒物质。
- 替代材料以减少对环境的影响。
- 材料有效的设计实践。这意味着尽量使用很少的材料。为了获得高强度结构的产品,有限元分析法和类似的系统能够用于优化产品的强度。只使用稀有的或昂贵的材料薄层,例如,在木纤维板或廉价基层上镀贵重金属。
- 替代性的复合材料有时能够减少建造过程中材料的总体数量,因为材料通常能够得到更好的利用。然而,循环混合材料通常更成问题,这些必须被认为违背了生产利益。
- 材料的选择和产品的设计应该考虑到将来的再循环和材料的使用。
- 制造交通工具时使用轻型材料以减少能源消耗。

产品材料的改变通常意味着改进其性能。所有材料具有其自身属性,如强度,或轻的重量,这更适用于产品。设计因此与材料的使用密切相关。

Free Reading

Sustainable Design

Sustainable design (also called environmental design, environmentally sustainable design, environmentally conscious design, etc.) is the philosophy of designing physical objects, the built environment, and services to comply with the principles of economic, social, and ecological sustainability.

Intentions

The intention of sustainable design is to "eliminate negative environmental impact completely through skillful, sensitive design". Manifestations of sustainable design require no non-renewable resources, impact the environment minimally, and relate people with the natural environment.

Applications

Applications of this philosophy range from the microcosm—small objects for everyday use, through to the macrocosm—buildings, cities, and the Earth's physical surface. It is a philosophy that can be applied in the fields of architecture, landscape architecture, urban design, urban planning, engineering, graphic design, industrial design, interior design, and fashion design.

Sustainable design is mostly a general reaction to global environmental crises, the rapid growth of economic activity and human population, depletion of natural resources, damage to ecosystems, and loss of biodiversity.

The limits of sustainable design are reducing. Whole earth impacts are beginning to be considered because growth in goods and services is consistently outpacing gains in efficiency. As a result, the net effect of sustainable design to date has been to simply improve the efficiency of rapidly increasing impacts. The present approach, which focuses on the efficiency of delivering individual goods and services, does not solve this problem. The basic dilemmas include: the increasing complexity of efficiency improvements; the difficulty of implementing new technologies in societies built around old ones; that physical impacts of delivering goods and services are not localized, but are distributed throughout the economies; and that the scale of resource use is growing and not stabilizing.

Sustainable Design Principles

While the practical application varies among disciplines, some common principles are as follows:

- Low-impact materials: choose non-toxic, sustainably produced or recycled materials which require little energy to process
- Energy efficiency: use manufacturing processes and produce products which require less energy
- Quality and durability: longer-lasting and better-functioning products will have to be replaced less frequently, reducing the impacts of producing replacements
- Design for reuse and recycling: "Products, processes, and systems should be designed for performance in a commercial 'afterlife'."
- Design impact measures for total carbon footprint and life-cycle assessment for any resource used are increasingly required and available. Many are complex, but some give quick and accurate whole-earth estimates of impacts. One measure estimates any spending as consuming an average economic share of global energy use of 8,000btu per dollar and producing CO_2 at the average rate of 0.57 kg of CO_2 per dollar (1995 US dollars) from DOE figures.
- Sustainable design standards and project design guides are also increasingly available and are vigorously being developed by a wide array of private organizations and individuals. There is also a large body of new methods emerging from the rapid development of what has become known as "sustainability science" promoted by a wide variety of educational and governmental institutions.
- Biomimicry: "redesigning industrial systems on biological lines … enabling the constant reuse of materials in continuous closed cycles…"
- Service substitution: shifting the mode of consumption from personal ownership of products to provision of services which provide similar functions, e.g., from a private automobile to a carsharing service. Such a system promotes minimal re-

- Renewability: materials should come from nearby (local or bioregional), sustainably managed renewable sources that can be composted when their usefulness has been exhausted.
- Healthy Buildings: sustainable building design aims to create buildings that are not harmful to their occupants or to the larger environment. An important emphasis is on indoor environmental quality, especially indoor air quality.
- Robust eco-design: robust design principles are applied to the design of a pollution sources.

Examples of Sustainable Design

Sustainable planning

Cohousing community illustrating greenspace preservation, tightly clustered housing, and parking on periphery, Ann Arbor, Michigan, 2003.

Urban planners that are interested in achieving sustainable development or sustainable cities use various design principles and techniques when designing cities and their infrastructure. These include Smart Growth theory, Transit-oriented development, sustainable urban infrastructure and New Urbanism. Smart Growth is an urban planning and transportation theory that concentrates growth in infill sites within the existing infrastructure of a city or town to avoid urban sprawl; and advocates compact, transit-oriented development, walkable, bicycle-friendly land use, including mixed-use development with a range of housing choices. Transit-oriented development attempts to maximise access to public transport and thereby reduce the need for private vehicles. Public transport is considered a form of sustainable urban infrastructure, which is a design approach which promotes protected areas, energy-efficient buildings, wildlife corridors and distributed, rather than centralized, power generation and waste water treatment. New Urbanism is more of a social and aesthetic urban design movement than a green one, but it does emphasize diversity of land use and population, as well as walkable communities which inherently reduce the need for automotive travel.

Both urban and rural planning can benefit from including sustainability as a central criterion when laying out roads, streets, buildings and other components of the built environment. Conventional planning practice often ignores or discounts the natural configuration of the land during the planning stages, potentially causing ecological damage such as the stagnation of streams, mudslides, soil erosion, flooding and pollution. Applying methods such as scientific modelling to planned building projects can draw attention to problems before construction begins, helping to minimize damage to the natural environment.

Cohousing is an approach to planning based on the idea of intentional communities. Such projects often prioritize common space over private space resulting in grouped structures that preserve more of the surrounding environment.

Watershed assessment of carrying capacity; estuary, riparian zone restoration and groundwater recharge for hydrologic cycle viability; and other opportunities and issues about Water and the environment show that the foundation of smart growth lies in the protection and preservation of water resources. The total amount of precipitation landing on the surface of a community becomes the supply for the inhabitants. This supply amount then dictates the carrying capacity—the potential population—as supported by the "water crop".

Sustainable architecture

Sustainable architecture is the design of sustainable buildings. Sustainable architecture attempts to reduce the collective environmental impacts during the production of building components, during the construction process, as well as during the lifecycle of the building (heating, electricity use, carpet cleaning etc.). This design practice emphasizes efficiency of heating and cooling systems; alternative energy sources such as solar hot water, appropriate building siting, reused or recycled building materials; on-site power generation—solar technology, ground source heat pumps, wind power; rainwater harvesting for gardening, washing and aquifer recharge; and on-site waste management such as green roofs that filter and control stormwater runoff. This requires close cooperation of the design team, the architects, the engineers, and the client at all project stages, from site selection, scheme formation, material selection and procurement, to project implementation.

Sustainable architects design with sustainable living in mind. Sustainable vs green design is the challenge that designs not only reflect healthy processes and uses but are powered by renewable energies and site specific resources. A test for sustainable design is—can the design function for its intended use without fossil fuel—unplugged. This challenge suggests architects and planners design solutions that can function without pollution rather than just reducing pollution. As technology progresses in architecture and design theories and as examples are built and tested, architects will soon be able to create not only passive, null-emission buildings, but rather be able to integrate the entire power system into the building design. In 2004 the 59 home housing communities, the Solar Settlement, and a 60,000 sq ft (5,600 m^2) integrated retail, commercial and residential building, the Sun Ship, were completed by architect Rolf Disch in Freiburg, Germany. The Solar Settlement is the first housing community world wide in which every home, all 59, produce a positive energy balance.

Sustainable Landscape and Garden Design

Sustainable landscape architecture is a category of sustainable design and energy-efficient landscaping concerned with the planning and design of outdoor space. Design techniques include planting trees to shade buildings from the sun or protect them from wind, using local materials, on-site composting and chipping to reduce green waste hauling, and also may involve using drought-resistant plantings in arid areas (xeriscaping) and buying stock from local growers to avoid energy use in transportation.

Sustainable graphic design considers the environmental impacts of graphic design products (such as packaging, printed materials, publications, etc.) throughout a life cycle that includes: raw material; transformation; manufacturing; transportation; use; and disposal. Techniques for sustainable graphic design include: reducing the amount of materials required for production; using paper and materials made with recycled, post-consumer waste; printing with low-VOC inks; and using production and distribution methods that require the least amount of transport.

Disposable Products

Detergents, newspapers and other disposable items can be designed to decompose, in the presence of air, water and common soil organisms. The current challenge in this area is to design such items in attractive colors, at costs as low as competing items. Since most such items end up in landfills, protected from air and water, the utility of such disposable products is debated.

Eco Fashion and Home Accessories

Creative designers and artists are perhaps the most inventive when it comes to upcycling or creating new products from old waste. A growing number of designers' upcycle waste materials such as car window glass and recycled ceramics, textile offcuts from upholstery companies, and even decommissioned fire hose to make belts and bags. Whilst accessories may seem trivial when pitted against green scientific breakthroughs; the ability of fashion and retail to influence and inspire consumer behaviour should not be underestimated. Eco design may also use bi-products of industry, reducing the amount of waste being dumped in landfill, or may harness new sustainable materials or production techniques e. g. fabric made from recycled PET plastic bottles or bamboo textiles.

参 考 文 献

[1] 何人可. 工业设计史 [M]. 北京：高等教育出版社，2004.
[2] 何人可，等. 工业设计专业英语 [M]. 北京：北京理工大学出版社，2012.
[3] 程能林. 工业设计概论 [M]. 北京：机械工业出版社，2011.
[4] 王明旨. 工业设计概论 [M]. 北京：高度教育出版社，2007.
[5] 王昭. 工业设计专业英语 [M]. 北京：中国轻工业出版社，2006.
[6] 戴力农，范希嘉，刘国余. 工业设计与艺术设计专业核心基础英语 [M]. 北京：机械工业出版社，2007.
[7] Alvin Toffler. The Third Wave [M]. New York：Random House，1984.
[8] 王受之. 世界现代设计史 [M]. 北京：中国青年出版社，2002.
[9] 江建民，毛荫秋，毛溪. 中英双语工业设计 [M]. 北京：中国建筑工业出版社，2009.
[10] http：//www. referenceforbusiness. com/encyclopedia/Con－Cos/Corporate－Identity. html
[11] http：//www. icsid. org/about/about. htm
[12] http：//www. idsa. org/
[13] http：//www. dmi. org/
[14] http：//www. ideo. com/us/
[15] http：//www. frogdesign. com
[16] https：//en. wikipedia. org/wiki/Museum＿of＿Modern＿Art

北京大学出版社教材书目

✧ 欢迎访问教学服务网站 www.pup6.com，免费查阅已出版教材的电子书(PDF版)、电子课件和相关教学资源。
✧ 欢迎征订投稿。联系方式：010-62750667，童编辑，13426433315@163.com，pup_6@163.com，欢迎联系。

序号	书　　　名	标准书号	主编	定价	出版日期
1	机械设计	978-7-5038-4448-5	郑　江，许　瑛	33	2007.8
2	机械设计	978-7-301-15699-5	吕　宏	32	2013.1
3	机械设计	978-7-301-17599-6	门艳忠	40	2010.8
4	机械设计	978-7-301-21139-7	王贤民，霍仕武	49	2014.1
5	机械设计	978-7-301-21742-9	师素娟，张秀花	48	2012.12
6	机械原理	978-7-301-11488-9	常治斌，张京辉	29	2008.6
7	机械原理	978-7-301-15425-0	王跃进	26	2013.9
8	机械原理	978-7-301-19088-3	郭宏亮，孙志宏	36	2011.6
9	机械原理	978-7-301-19429-4	杨松华	34	2011.8
10	机械设计基础	978-7-5038-4444-2	曲玉峰，关晓平	27	2008.1
11	机械设计基础	978-7-301-22011-5	苗淑杰，刘喜平	49	2015.8
12	机械设计基础	978-7-301-22957-6	朱　玉	38	2014.12
13	机械设计课程设计	978-7-301-12357-7	许　瑛	35	2012.7
14	机械设计课程设计	978-7-301-18894-1	王　慧，吕　宏	30	2014.1
15	机械设计辅导与习题解答	978-7-301-23291-0	王　慧，吕　宏	26	2013.12
16	机械原理、机械设计学习指导与综合强化	978-7-301-23195-1	张占国	63	2014.1
17	机电一体化课程设计指导书	978-7-301-19736-3	王金娥　罗生梅	35	2013.5
18	机械工程专业毕业设计指导书	978-7-301-18805-7	张黎骅，吕小荣	22	2015.4
19	机械创新设计	978-7-301-12403-1	丛晓霞	32	2012.8
20	机械系统设计	978-7-301-20847-2	孙月华	32	2012.7
21	机械设计基础实验及机构创新设计	978-7-301-20653-9	邹旻	28	2014.1
22	TRIZ 理论机械创新设计工程训练教程	978-7-301-18945-0	蒯苏苏，马履中	45	2011.6
23	TRIZ 理论及应用	978-7-301-19390-7	刘训涛，曹　贺等	35	2013.7
24	创新的方法——TRIZ 理论概述	978-7-301-19453-9	沈萌红	28	2011.9
25	机械工程基础	978-7-301-21853-2	潘玉良，周建军	34	2013.2
26	机械工程实训	978-7-301-26114-9	侯书林，张　炜等	52	2015.10
27	机械 CAD 基础	978-7-301-20023-0	徐云杰	34	2012.2
28	AutoCAD 工程制图	978-7-5038-4446-9	杨巧绒，张克义	20	2011.4
29	AutoCAD 工程制图	978-7-301-21419-0	刘善淑，胡爱萍	38	2015.2
30	工程制图	978-7-5038-4442-6	戴立玲，杨世平	27	2012.2
31	工程制图	978-7-301-19428-7	孙晓娟，徐丽娟	30	2012.5
32	工程制图习题集	978-7-5038-4443-4	杨世平，戴立玲	20	2008.1
33	机械制图(机类)	978-7-301-12171-9	张绍群，孙晓娟	32	2009.1
34	机械制图习题集(机类)	978-7-301-12172-6	张绍群，王慧敏	29	2007.8
35	机械制图(第 2 版)	978-7-301-19332-7	孙晓娟，王慧敏	38	2014.1
36	机械制图	978-7-301-21480-0	李凤云，张　凯等	36	2013.1
37	机械制图习题集(第 2 版)	978-7-301-19370-7	孙晓娟，王慧敏	22	2011.8
38	机械制图	978-7-301-21138-0	张　艳，杨晨升	37	2012.8
39	机械制图习题集	978-7-301-21339-1	张　艳，杨晨升	24	2012.10
40	机械制图	978-7-301-22896-8	臧福伦，杨晓冬等	60	2013.8
41	机械制图与 AutoCAD 基础教程	978-7-301-13122-0	张爱梅	35	2013.1
42	机械制图与 AutoCAD 基础教程习题集	978-7-301-13120-6	鲁　杰，张爱梅	22	2013.1
43	AutoCAD 2008 工程绘图	978-7-301-14478-7	赵润平，宗荣珍	35	2009.4
44	AutoCAD 实例绘图教程	978-7-301-20764-2	李庆华，刘晓杰	32	2012.6
45	工程制图案例教程	978-7-301-15369-7	宗荣珍	28	2009.6
46	工程制图案例教程习题集	978-7-301-15285-0	宗荣珍	24	2009.6
47	理论力学(第 2 版)	978-7-301-23125-8	盛冬发，刘　军	38	2013.9
48	材料力学	978-7-301-14462-6	陈忠安，王　静	30	2013.4
49	工程力学(上册)	978-7-301-11487-2	毕勤胜，李纪刚	29	2008.6
50	工程力学(下册)	978-7-301-11565-7	毕勤胜，李纪刚	28	2008.6
51	液压传动(第 2 版)	978-7-301-19507-9	王守城，容一鸣	38	2013.7
52	液压与气压传动	978-7-301-13179-4	王守城，容一鸣	32	2013.7

序号	书名	标准书号	主编	定价	出版日期
53	液压与液力传动	978-7-301-17579-8	周长城等	34	2011.11
54	液压传动与控制实用技术	978-7-301-15647-6	刘忠	36	2009.8
55	金工实习指导教程	978-7-301-21885-3	周哲波	30	2014.1
56	工程训练(第3版)	978-7-301-24115-8	郭永环，姜银方	38	2016.1
57	机械制造基础实习教程	978-7-301-15848-7	邱兵，杨明金	34	2010.2
58	公差与测量技术	978-7-301-15455-7	孔晓玲	25	2012.9
59	互换性与测量技术基础(第3版)	978-7-301-25770-8	王长春等	35	2015.6
60	互换性与技术测量	978-7-301-20848-9	周哲波	35	2012.6
61	机械制造技术基础	978-7-301-14474-9	张鹏，孙有亮	28	2011.6
62	机械制造技术基础	978-7-301-16284-2	侯书林 张建国	32	2012.8
63	机械制造技术基础	978-7-301-22010-8	李菊丽，何绍华	42	2014.1
64	先进制造技术基础	978-7-301-15499-1	冯宪章	30	2011.11
65	先进制造技术	978-7-301-22283-6	朱林，杨春杰	30	2013.4
66	先进制造技术	978-7-301-20914-1	刘璇，冯凭	28	2012.8
67	先进制造与工程仿真技术	978-7-301-22541-7	李彬	35	2013.5
68	机械精度设计与测量技术	978-7-301-13580-8	于峰	25	2013.7
69	机械制造工艺学	978-7-301-13758-1	郭艳玲，李彦蓉	30	2008.8
70	机械制造工艺学(第2版)	978-7-301-23726-7	陈红霞	45	2014.1
71	机械制造工艺学	978-7-301-19903-9	周哲波，姜志明	49	2012.1
72	机械制造基础(上)——工程材料及热加工工艺基础(第2版)	978-7-301-18474-5	侯书林，朱海	40	2013.2
73	制造之用	978-7-301-23527-0	王中任	30	2013.12
74	机械制造基础(下)——机械加工工艺基础(第2版)	978-7-301-18638-1	侯书林，朱海	32	2012.5
75	金属材料及工艺	978-7-301-19522-2	于文强	44	2013.2
76	金属工艺学	978-7-301-21082-6	侯书林，于文强	32	2012.8
77	工程材料及其成形技术基础(第2版)	978-7-301-22367-3	申荣华	58	2016.1
78	工程材料及其成形技术基础学习指导与习题详解(第2版)	978-7-301-26300-6	申荣华	28	2015.9
79	机械工程材料及成形基础	978-7-301-15433-5	侯俊英，王兴源	30	2012.5
80	机械工程材料(第2版)	978-7-301-22552-3	戈晓岚，招玉春	36	2013.6
81	机械工程材料	978-7-301-18522-3	张铁军	36	2012.5
82	工程材料与机械制造基础	978-7-301-15899-9	苏子林	32	2011.5
83	控制工程基础	978-7-301-12169-6	杨振中，韩致信	29	2007.8
84	机械制造装备设计	978-7-301-23869-1	宋士刚，黄华	40	2014.12
85	机械工程控制基础	978-7-301-12354-6	韩致信	25	2008.1
86	机电工程专业英语(第2版)	978-7-301-16518-8	朱林	24	2013.7
87	机械制造专业英语	978-7-301-21319-3	王中任	28	2014.12
88	机械工程专业英语	978-7-301-23173-9	余兴波，姜波等	30	2013.9
89	机床电气控制技术	978-7-5038-4433-7	张万奎	26	2007.9
90	机床数控技术(第2版)	978-7-301-16519-5	杜国臣，王士军	35	2014.1
91	自动化制造系统	978-7-301-21026-0	辛宗生，魏国丰	37	2014.1
92	数控机床与编程	978-7-301-15900-2	张洪江，侯书林	25	2012.10
93	数控铣床编程与操作	978-7-301-21347-6	王志斌	35	2012.10
94	数控技术	978-7-301-21144-1	吴瑞明	28	2012.9
95	数控技术	978-7-301-22073-3	唐友亮 余勃	45	2014.1
96	数控技术(双语教学版)	978-7-301-27920-5	吴瑞明	36	2017.3
97	数控技术与编程	978-7-301-26028-9	程广振 卢建湘	36	2015.8
98	数控技术及应用	978-7-301-23262-0	刘军	49	2013.10
99	数控加工技术	978-7-5038-4450-7	王彪，张兰	29	2011.7
100	数控加工与编程技术	978-7-301-18475-2	李体仁	34	2012.5
101	数控编程与加工实习教程	978-7-301-17387-9	张春雨，于雷	37	2011.9
102	数控加工技术及实训	978-7-301-19508-6	姜永成，夏广岚	33	2011.9
103	数控编程与操作	978-7-301-20903-5	李英平	26	2012.8
104	数控技术及其应用	978-7-301-27034-9	贾伟杰	40	2016.4
105	现代数控机床调试及维护	978-7-301-18033-4	邓三鹏等	32	2010.11
106	金属切削原理与刀具	978-7-5038-4447-7	陈锡渠，彭晓南	29	2012.5
107	金属切削机床(第2版)	978-7-301-25202-4	夏广岚，姜永成	42	2015.1
108	典型零件工艺设计	978-7-301-21013-0	白翔清	34	2012.8
109	模具设计与制造(第2版)	978-7-301-24801-0	田光辉，林红旗	56	2016.1
110	工程机械检测与维修	978-7-301-21185-4	卢彦群	45	2012.9

序号	书 名	标准书号	主 编	定价	出版日期
111	工程机械电气与电子控制	978-7-301-26868-1	钱宏琦	54	2016.3
112	工程机械设计	978-7-301-27334-0	陈海虹，唐绪文	49	2016.8
113	特种加工(第2版)	978-7-301-27285-5	刘志东	54	2017.3
114	精密与特种加工技术	978-7-301-12167-2	袁根福，祝锡晶	29	2011.12
115	逆向建模技术与产品创新设计	978-7-301-15670-4	张学昌	28	2013.1
116	CAD/CAM 技术基础	978-7-301-17742-6	刘 军	28	2012.5
117	CAD/CAM 技术案例教程	978-7-301-17732-7	汤修映	42	2010.9
118	Pro/ENGINEER Wildfire 2.0 实用教程	978-7-5038-4437-X	黄卫东，任国栋	32	2007.7
119	Pro/ENGINEER Wildfire 3.0 实例教程	978-7-301-12359-1	张选民	45	2008.2
120	Pro/ENGINEER Wildfire 3.0 曲面设计实例教程	978-7-301-13182-4	张选民	45	2008.2
121	Pro/ENGINEER Wildfire 5.0 实用教程	978-7-301-16841-7	黄卫东，郝用兴	43	2014.1
122	Pro/ENGINEER Wildfire 5.0 实例教程	978-7-301-20133-6	张选民，徐超辉	52	2012.2
123	SolidWorks 三维建模及实例教程	978-7-301-15149-5	上官林建	30	2012.8
124	UG NX 9.0 计算机辅助设计与制造实用教程(第2版)	978-7-301-26029-6	张黎骅，吕小荣	36	2015.8
125	CATIA 实例应用教程	978-7-301-23037-4	于志新	45	2013.8
126	Cimatron E9.0 产品设计与数控自动编程技术	978-7-301-17802-7	孙树峰	36	2010.9
127	Mastercam 数控加工案例教程	978-7-301-19315-0	刘 文，姜永梅	45	2011.8
128	应用创造学	978-7-301-17533-0	王版军，沈豫浙	26	2012.5
129	机电产品学	978-7-301-15579-0	张亮峰等	24	2015.4
130	品质工程学基础	978-7-301-16745-8	丁 燕	30	2011.5
131	设计心理学	978-7-301-11567-1	张成忠	48	2011.6
132	计算机辅助设计与制造	978-7-5038-4439-6	仲梁维，张国全	29	2007.9
133	产品造型计算机辅助设计	978-7-5038-4474-4	张慧姝，刘永翔	27	2006.8
134	产品设计原理	978-7-301-12355-3	刘美华	30	2008.2
135	产品设计表现技法	978-7-301-15434-2	张慧姝	42	2012.5
136	CorelDRAW X5 经典案例教程解析	978-7-301-21950-8	杜秋磊	40	2013.1
137	产品创意设计	978-7-301-17977-2	虞世鸣	38	2012.5
138	工业产品造型设计	978-7-301-18313-7	袁涛	39	2011.1
139	化工工艺学	978-7-301-15283-6	邓建强	42	2013.7
140	构成设计	978-7-301-21466-4	袁涛	58	2013.1
141	设计色彩	978-7-301-24246-9	姜晓微	52	2014.6
142	过程装备机械基础(第2版)	978-301-22627-8	于新奇	38	2013.7
143	过程装备测试技术	978-7-301-17290-2	王毅	45	2010.6
144	过程控制装置及系统设计	978-7-301-17635-1	张早校	30	2010.8
145	质量管理与工程	978-7-301-15643-8	陈宝江	34	2009.8
146	质量管理统计技术	978-7-301-16465-5	周友苏，杨 飒	30	2010.1
147	人因工程	978-7-301-19291-7	马如忠	39	2011.8
148	工程系统概论——系统论在工程技术中的应用	978-7-301-17142-4	黄志坚	32	2010.6
149	测试技术基础(第2版)	978-7-301-16530-0	江征风	30	2014.1
150	测试技术实验教程	978-7-301-13489-4	封士彩	22	2008.8
151	测控系统原理设计	978-7-301-24399-2	齐永奇	39	2014.7
152	测试技术学习指导与习题详解	978-7-301-14457-2	封士彩	34	2009.3
153	可编程控制器原理与应用(第2版)	978-7-301-16922-3	赵 燕，周新建	33	2011.11
154	工程光学	978-7-301-15629-2	王红敏	28	2012.5
155	精密机械设计	978-7-301-16947-6	田 明，冯进良等	38	2011.9
156	传感器原理及应用	978-7-301-16503-4	赵 燕	35	2014.1
157	测控技术与仪器专业导论(第2版)	978-7-301-24223-0	陈毅静	36	2014.6
158	现代测试技术	978-7-301-19316-7	陈科山，王 燕	43	2011.8
159	风力发电原理	978-7-301-19631-1	吴双群，赵丹平	33	2011.10
160	风力机空气动力学	978-7-301-19555-0	吴双群	32	2011.10
161	风力机设计理论及方法	978-7-301-20006-3	赵丹平	32	2012.1
162	计算机辅助工程	978-7-301-22977-4	许承东	38	2013.8
163	现代船舶建造技术	978-7-301-23703-8	初冠南，孙清洁	33	2014.1
164	机床数控技术(第3版)	978-7-301-24452-4	杜国臣	43	2016.8
165	机械设计课程设计	978-7-301-27844-4	王 慧，吕 宏	36	2016.12
166	工业设计概论(双语)	978-7-301-27933-5	窦金花	35	2017.3
167	产品创新设计与制造教程	978-7-301-27921-2	赵 波	31	2017.3

如您需要免费纸质样书用于教学，欢迎登陆第六事业部门户网(www.pup6.com)填表申请，并欢迎在线登记选题以到北京大学出版社来出版您的大作，也可下载相关表格填写后发到我们的邮箱，我们将及时与您取得联系并做好全方位的服务。